The Origins and Development of Roger J. Williams'
Concept of Biochemical Individuality

Medizingeschichte im Kontext

Edited by Karl-Heinz Leven, Mariacarla Gadebusch
Bondio, Hans-Georg Hofer and Livia Prüll

Founded as Freiburger Forschungen zur Medizingeschichte
by Ludwig Aschoff, continued by Eduard Seidler

Band 26

Georg-Benedict Brand

The Origins and Development of Roger J. Williams' Concept of Biochemical Individuality

Berlin · Bruxelles · Chennai · Lausanne · New York · Oxford

Library of Congress Cataloging-in-Publication Data
A CIP catalog record for this book has been applied for at the Library of Congress.

Bibliographic Information published by the Deutsche Nationalbibliothek
The Deutsche Nationalbibliothek lists this publication in the Deutsche Nationalbibliografie; detailed bibliographic data is available online at http://dnb.d-nb.de.

DE-5-41
ISSN 1437-3122
ISBN 978-3-631-92772-4 (Print)
E-ISBN 978-3-631-92773-1 (E-PDF)
E-ISBN 978-3-631-92774-8 (E-PUB)
DOI 10.3726/b22420

© 2024 Peter Lang Group AG, Lausanne
Published by Peter Lang GmbH, Berlin, Germany

info@peterlang.com - www.peterlang.com

All rights reserved.

All parts of this publication are protected by copyright. Any utilisation outside the strict limits of the copyright law, without the permission of the publisher, is forbidden and liable to prosecution. This applies in particular to reproductions, translations, microfilming, and storage and processing in electronic retrieval systems.

Written with authorisation of the
Faculty of Medicine of the University of Bonn

First reviewer:* Prof. Dr. phil. Dr. rer. med. habil. Mariacarla Gadebusch Bondio
Second reviewer: Prof. Dr. Hans-Georg Hofer

Day of oral examination: 22.07.2024

From the Institute for Medical Humanities
Director: Prof. Dr. phil. Dr. rer. med. habil. Mariacarla Gadebusch Bondio

To my family

"Human worth resides not only in those whom we regard as great, but in all of us, and we should provide an environment which will give everyone an equal chance to develop his potentialities in the way best suited to him individually."

- Roger J. Williams, Free and Unequal (1953)

Table of contents

List of abbreviations 13

1 Introduction 15
   1.1 Why Individuality? 15
   1.2 Literature Review 17
   1.3 Scope and Research Questions 21

2 Materials and Methods 23
   2.1 Source Discussion and Research Methodology 23

3 A Career in Biochemistry 29
   3.1 Upbringing and Education in Biochemistry 29
   3.2 Academic Networks 31

4 Origins of Biochemical Individuality as a Concept 39
   4.1 Youthful Observations 39
   4.2 "The Vitamine Requirement of Yeast" 40
   4.3 Adverse Drug Reaction 42
   4.4 Yeast Extract Research 44
   4.5 "'Taste Deficiency' for Creatine" 47
   4.6 Biochemical Individuality Before 1940 50
   4.7 Conclusion 50

5 Vitamin Studies 51
   5.1 Individuality of Yeast Strains 52
   5.2 The Vitamin Content of Tissues 54
   5.3 Practical Applications of Vitamin Research 59
   5.4 *What to Do About Vitamins* 65

5.5 Biochemical Individuality Preceding *The Human Frontier* 69
5.6 Conclusion 69

6 The Human Frontier 71
6.1 "Humanity Must Understand Itself" 72
6.2 Audience 74
6.3 Social Control 88
6.4 "Distinctive Metabolic Traits" 90
6.5 "Individuals Vary Greatly" 96
6.6 Reviews of *The Human Frontier* 102
6.7 Biochemical Individuality Following *The Human Frontier* 108
6.8 Conclusion 108

7 The Scientific Study of Individuals and Alcoholism 109
7.1 "Biochemical Individuality and Its Implications" 109
7.2 "The Etiology of Alcoholism" 127
7.3 "Alcoholics and Metabolism" 130
7.4 *An Introduction to Biochemistry*, Second Edition 132
7.5 Anti-Communist Sentiment 137
7.6 Symposium and Society 139
7.7 Biochemical Individuality in 1949 144
7.8 Conclusion 144

8 Genetotrophic Disease 145
8.1 The Metabolic Individualities of Rats 145
8.2 Collaborative Individuality 150
8.3 Genetotrophic Promotion 152
8.4 Biochemical Individuality Following the Genetotrophic Principle 153
8.5 Conclusion 154

Table of contents

9 Human Individuality 155
   9.1 Biochemical Individuality V 155
   9.2 Of Marbles and Men 157
   9.3 *Nutrition and Alcoholism* 159
      9.3.1 Evidence 160
      9.3.2 Supplementation 162
      9.3.3 Appeal 163
      9.3.4 Promotion and Reviews of *Nutrition and Alcoholism* 164
      9.3.5 Summary 167
   9.4 Biochemical Institute Studies 167
   9.5 Biochemical Individuality Following First Human Research 170
   9.6 Conclusion 170

10 Free and Unequal 171
   10.1 Simple Yet Profound 171
   10.2 Signatures 173
   10.3 Politicisation 173
      10.3.1 Communism and Racism Revisited 174
      10.3.2 Assembly Line Educations, Regimentation, and Dogma 181
   10.4 Reviews of *Free and Unequal* 183
   10.5 Biochemical Individuality Following *Free and Unequal* 184
   10.6 Conclusion 185

11 Practical Genetotrophism 187
   11.1 Genetotrophic Supplementation 187
   11.2 Cancer 188
   11.3 Individual Anatomies and Compositions 189
   11.4 Chemical Anthropology 192
   11.5 Normal Young Men 193

- 11.6 The Concept of Biochemical Individuality Before *Biochemical Individuality* 194
- 11.7 Conclusion 194

12 Biochemical Individuality 195
- 12.1 Evidence 203
  - 12.1.1 From Basic Genetics to All-Encompassing Variation 206
- 12.2 Variation and Its Significance 207
- 12.3 Resistance and Publishing Difficulties 210
- 12.4 Reviews of *Biochemical Individuality* 211
- 12.5 Appreciation Through the Years 214
- 12.6 Biochemical Individuality in Its Final Form 214
- 12.7 Conclusion 214

13 Results 215

14 Discussion 221

15 Summary 235

16 Annex 239
- 16.1 Interview with Donald R. Davis 239

17 List of figures 245

18 List of tables 249

19 References 251
- 19.1 Unpublished Materials 251
- 19.2 Published Materials 260

20 Index 287

21 Acknowledgments 289

List of abbreviations

BI	Biochemical Individuality
Biochemical Individuality	*Biochemical Individuality: The Basis for the Genetotrophic Concept* (Williams, 1956a)
DBUT	The Department of Biochemistry at The University of Texas at Austin
EBM	Evidence-Based Medicine
Free and Unequal	*Free and Unequal: The Biological Basis of Individual Liberty* (Williams, 1953a)
GC	The Genetotrophic Concept / The Genetotrophic Principle
JAMA	*The Journal of the American Medical Association*
MI	Metabolic Idiosyncrasy
OSC	Oregon State College
PM	Personalised Medicine
PTC	Phenyl Thiocarbamide
The Human Frontier	*The Human Frontier: A New Pathway for Science Toward a Better Understanding of Ourselves* (Williams, 1946a)
USSR	Union of Socialist Soviet Republics
UT	The University of Texas at Austin

1 Introduction

1.1 Why Individuality?

Since work on this dissertation commenced in 2018, a new generation, the bid for alphabetical continuity bestowing upon them the label "Z", has joined the world's adult population. This generation of "Body-Positivity/-Neutrality"[1] and "BeReal."[2] challenges a wide variety of societal norms (Clarke, 2022; Cowles, 2022; Hawley, 2022). Central to this generational revolution is a more pronounced appreciation for the importance of every individuals' distinctiveness, followed by a plea for increased attention to the respective needs this variation creates (Chillakuri and Mahanandia, 2018; Francis and Hoefel, 2018; Schenarts, 2020). Older generations are often nonplussed by the bravado with which "Gen Z" shuns labels and norms in pursuit of their own truth and happiness (Francis and Hoefel, 2018). Alongside these societal and cultural changes, medicine and science are also becoming increasingly appreciative of the individuality inherent to all living organisms. With standard medical practices being changed to incorporate this understanding, most recently through a policy change by the American Medical Association regarding the Body Mass Index (American Medical Association, 2023; Blum, 2023), personalised and tailored treatments are becoming more affordable and widespread. The growing number of medical literatures discussing the ramifications and potentials of personalised medicine additionally indicates the growing relevance of human individuality within the medical field.[3] The increased attention to this inherent uniqueness which

---

1 The movement of "Body-Positivity" challenges previously held ideals of beauty with the goal of normalising all body shapes and sizes, aiming to foster positive body images throughout society (Cohen et al., 2021). Conversely, "Body-Neutrality" does not necessitate positive feelings regarding an individual's body or weight, emphasising the acceptance of its individuality and focusing on the individual's health and happiness (Cowles, 2022).
2 "BeReal." is a social media app based on the sharing of real-time images within a defined daily 2-minute time limit. Its aim is to provide "a new and unique way to discover who your friends really are in their daily life" (Perreau, 2023). It has become widely popular with members of "Generation Z" and is largely seen as a more authentic alternative to other forms of social media, as it does not provide the option of editing or filtering the posted images (Clarke, 2022; Hawley, 2022).
3 See, for example, Cesario et al. (2023) and Jain (2021).

all human beings encompass means understanding the roots of our concepts of individuality has become more relevant than ever.

There has long been a certain natural awareness of interindividual variation between human beings. This notion is reflected in all manner of historical allusions to individuality in science and society, which can be dated as far back as the ancient Greek philosophers and their medical principles (Galen, 2012; Hippokrates, 2014, pp. 108–109, 364). The nature of these descriptions is often nebulous, indicating a widespread variation inherent to all humans and animals. In accordance with the scientific theories and practices of the time, these early notions do not seek to categorically and scientifically collect and process evidence of individuality, appreciating variation as an unalterable fact of life and thereby also as a factor in disease. Their portrayals are often based on macroscopic differences; the anatomical differences with regard to size, colour, shape, and symmetry of the human body being marked and providing the most palpable evidence for interindividual variation. Though such discrepancies on a macroscopic level still bear relevance to modern concepts of individuality, contemporary research largely focuses on deviations in cellular, metabolic, and enzymic composition (Jain, 2021, pp. 5–17). The relevance and importance of such historic viewpoints is widely discussed within scientific literature on the development of concepts pertaining to human individuality (Gadebusch Bondio, 2015, pp. 20–25; Gadebusch Bondio and Spöring, 2017 pp. 38–39; Jain, 2009, p. 4).

The work of Roger J. Williams (1893–1988) brought the true extent of this variation, which is often microscopic and intangible in nature, to light. While scientists and medics of previous generations had discussed and recognised certain aspects of a human discrepancies (Galen, 2012; Garrod, 1902, 1923; Hippokrates, 2014, Chap. 1),[4] Williams began, even adamantly demanded, the systematic study of the individuality shown by human beings. A biochemist by training, he initially made a name for himself through the discovery of pantothenic acid and the study of B vitamins in the 1920s and 1930s (Davis et al., 2008). His professional interest in concepts of individuality developed while engaged in this fundamental biochemical research (Davis et al., 2008; Williams, 1931). Noting the fact that "an interest in variations and individuality had often been considered a hobby and has not led to serious publications", and that the study of human differences had historically "not gained the respectability that it deserves", Williams' research and collection of data structured and sophisticated our knowledge of Biochemical Individuality (BI) (Williams, 1956a, p. X).[5]

---

4 For further reading, see Gadebusch Bondio (2015).
5 "Biochemical Individuality" is also referred to as "BI" within this thesis.

In his ground-breaking book, *Biochemical Individuality: The Basis for the Genetotrophic Concept* (*Biochemical Individuality*),[6] Roger J. Williams presents vast amounts of scientific evidence of human variation and proposes his concept of what he calls Biochemical Individuality. Based on decades of research and experience in biochemistry, his concept unifies genetic and biochemical theories, suggesting that every human being is entirely unique in all possible aspects of their fundamental (bio-)chemical composition. Published in 1956, this book is largely considered a standard work on human variation.[7] Its relevance still appreciated more than a quarter of a century later, it was reissued in the late 1990s and is still available for purchase today (Williams, 1998).

1.2 Literature Review

The available research discussing Williams' personal and professional vita can be divided into three categories. The largest category of publications depicting aspects of Williams' life and work are biographical analyses. A second class comprises works on medical history and epistemology. The third and final category encompasses discussions of Williams' work in contemporary medical and scientific literature. Though his research remains relevant to our ideas of uniqueness and individualism until this day, Williams' oeuvre and his development of BI have not previously been analysed comprehensively.[8]

The first and largest category to be discussed are the short biographies and memoirs published in the wake of Williams' death in 1988 (Biesele, 1988; Biochemist Roger Williams dead at 94, 1988; Davis, 2010, 2003a, 1988; Davis et al., 2008; Fowler, 1988). Analogously, an official document of the city of Austin provides a short overview of his life and career (City of Austin, 2009). These works discuss his professional and personal successes in varying detail, generally not delving into the specific aspects of his work, with most focusing on his vitamin research. The most detailed biographical piece has been published by Donald R. Davis, Marvin L. Hackert, and Lester J. Reed; three biochemists and former colleagues of Roger Williams at The University of Texas at Austin (UT).[9] Their biographical memoir, published by the National Academy of Science on

---

6 "*Biochemical Individuality: The Basis for the Genetotrophic Concept* (Williams, 1956a)" is also referred to as "*Biochemical Individuality*" within this thesis.
7 See Section 1.2. Literature Review.
8 Large proportions of evidence have previously been unpublished and are, therefore, not analysed in the works discussed below.
9 "The University of Texas at Austin" is also referred to as "UT" within this thesis.

the twentieth anniversary of Williams' death in 2008, is the most holistic eulogy, depicting his private life and entire academic career (Davis et al., 2008). It reflects on the personal facts presented in Williams' autobiography, on which it is heavily based, and adds the insight of scientific contemporaries regarding his professional work. Despite the personal attachment of the three authors to Williams and his (and in some instances their own) work and the self-stylisation by Williams in his autobiography, their biography paints an accurate picture of his life and career. This biographical overview provides a framework within which Williams' research can be understood and has been invaluable in forming an overview of his personal and professional existence.

A further category of publications to be discussed is the historical and epistemological research examining Williams work and his theories. The 1998 reissue of *Biochemical Individuality* includes a foreword by Jeffrey Bland, appraising the importance and continued relevance of Williams' theories (Williams, 1998, Foreword). Geneticist, colleague, and contemporary Arno Motulsky (1923–2018) discusses Williams' concept and its relevance to the development of Pharmacogenomics (Motulsky, 2002).[10] Speaking in critical tones of his research, Motulsky presents Williams' work as a part of the greater development of the field while presenting a scientific appraisal of the work of his colleagues and predecessors. Other contemporary critiques and comments on the research of the Department of Biochemistry at The University of Texas under Roger Williams' leadership indicate the reception of his research and concepts and will be discussed in relation to the works they critique.[11] These are, therefore, omitted from this literature review.

In a chapter on "The Epistemics of 'Personalised Medicine'", Williams' work is presented within the context of a developing personalised medicine by Michl,[12] including a representation of his research and its consequences (Michl, 2015, pp. 70–74). Characterising Williams' work as a combination of genetic,

---

10   For further information, see Gartler and Jarvik (2019).
11   "The Department of Biochemistry at The University of Texas at Austin" is also referred to as "DBUT" within this thesis.
12   "Personalised Medicine" is also referred to as "PM" within this thesis. Various definitions of this term exist, with none accepted universally. The European Commission defines PM as "A medical model using characterization of individuals' phenotypes and genotypes (e.g. molecular profiling, medical imaging, lifestyle data) for tailoring the right therapeutic strategy for the right person at the right time, and/or to determine the predisposition to disease and/or to deliver timely and targeted prevention" (European Commission, 2023).

environmental, and biochemical research, the ingenuity of his approach to the study of humankind is highlighted (Michl, 2015, p. 71). Distinguishing himself from prior models of individuality, which largely involved the categorisation of human beings into "normal" and "abnormal", Williams' terminological dissociation, and the methodological consequences which follow in his research, are deliberated. Furthermore, Michl (2015, p. 73) considers the ramifications of Williams' research outside of purely scientific matters, accentuating his efforts to address the general population and non-scientific societal factions. This work additionally comprises an analysis of the methodology behind Williams' concept of BI and therefore provides an effective overview of the concept central to this thesis.

Similarly, Gadebusch Bondio reflects on the work of Roger Williams on multiple occasions. In work on "Beyond the Causes of Disease: Prediction and the Need for a New Philosophy of Medicine and its Historical Development" in *Medical Ethics, Prediction and Prognosis*, the first section is dedicated to Williams and his ideas (Gadebusch Bondio, 2017, Chap. 1). Delineating Williams' Genetotrophic Concept and his concept of Biochemical Individuality, this analysis describes the importance of Williams' concepts to contemporary ideas of individuality in disease, while equating the consequence of his research to the works of other pioneering scientists of the 20th century (Gadebusch Bondio, 2017, pp. 12–16). Williams' concepts are here depicted as a central stepping-stone in the development of personalised and predictive medicine, while the critique of his work by contemporaries of other scientific fields is additionally considered. A further piece depicting a historical timeline of the development of personalised medicine additionally refers to Williams and his work, though the focus of the analysis lies on the work of Theodor Brugsch (1878–1963) and his "Personallehre" of the early 20th century (Gadebusch Bondio and Spöring, 2017, p. 43). Here, the concepts set forth in *Biochemical Individuality* and Williams' research on alcoholism are portrayed alongside the early works of American pharmacogeneticists as a continuance of Brugsch's concepts and theories (Gadebusch Bondio and Spöring, 2017, pp. 43–46). Williams' work is further discussed in an article on the historical development of the concept of human individuality in medicine (Gadebusch Bondio, 2015). Embedding his investigations within a discussion of the historical scientific unease regarding individuals in research, Williams' work is again depicted as a central scientific stepping stone to current ideas of individuality. Considering Williams' research from the viewpoint of naturopathic and integrative medicine, Pizzorno (2019) provides a further short overview of his career and illustrates Williams' importance to his own views on medical treatment and research. These works contextualise Williams' research within the framework of his contemporary scientific community, relating his studies to the individualised medicine of today. Williams is conspicuous as a biochemist in such literature on the history of personalised medicine, often placed alongside a relatively homogenous group of human geneticists and described as a unifying force between biochemistry and genetics (Michl, 2015, p. 66). The information portrayed by his theories, as presented in *Biochemical Individuality*, has considerable correspondences with the fundamental principles upon which personalised medicine is founded today, and can therefore help understand these philosophies further. Such historical and epistemological research helps to reveal the continued relevance of Williams' investigations and provides comparative information regarding Williams' contemporaries and the research which preceded him.

The continued applicability of Williams' work is indicated by medical and scientific contributions in which he is cited as the originator of our concepts of individuality today (Gonzalez and Massari, 2012, p. 2; Neustadt and Pieczenik, 2007, p. 30). Within the field of nutrition, Williams' research is described most favourably, with his Genetotrophic Concept and Biochemical Individuality seen as essential forerunners to contemporary nutritional principles (Bland, 2019a, 2019b; Gasta, 2020; Johnson and Hand, 2020; de Las Hazas and Dávalos, 2022; Steg et al., 2022). Schloss (2023) additionally indicates how Williams' nutritional research in connection with Biochemical Individuality may hold answers regarding susceptibility to long COVID.[13] Similarly, Williams' work on BI is positively discussed as an inspiration within integrative medicine (Fitzgerald and Rountree, 2022), as well as his research being described as a driver for biochemistry in the 20th century (Giera et al., 2022). Badrick (2021) further indicates Williams' importance to clinical chemist's understanding of variation, while Patterson and Turnbaugh (2014) indicate how Williams' concept shaped the development of microbial research in pharmacology and toxicology. Williams' work is not subject to in-depth discussion in this contemporary scientific literature, these works citing his research as a natural forerunner to modern concepts of individuality and shortly summarising the essence of BI as set forth in *Biochemical Individuality* in 1956, while accepting these without question. The research of other scientists in this context is not mentioned or discussed in the aforementioned works, with Williams' publication and theories portrayed as a pioneering

---

13 Long COVID is also referred to as Post-COVID Syndrome and describes a wide variety of ongoing symptoms following an infection with the COVID-19 virus. For further information, see National Health Service (2023).

historical foundation on which further research can be based. Williams' ideas therefore remain highly relevant to current medical and scientific research, providing the groundwork for modern techniques and ideas.

1.3 Scope and Research Questions

The available literature offers an effective overview of Williams' life and work, discussing multiple aspects of his theories. Portraying the concept of BI in its final form, these discussions create the impression of a singular theory, postulated in 1956 alone, as the intricacies of the development of his ideas are not deliberated. The complex advances of Biochemical Individuality, from unformed, vague ideas of variation to a concrete and many-layered theory encompassing all of science and society, therefore form the basis of discussion within this thesis. Additionally, the available publications largely focus on Williams' work within a specific scientific or historical context; a holistic discussion of the outside influences upon Williams and his work are therefore also examined here. This analysis seeks to elucidate these facets of Williams' research and the concept of Biochemical Individuality, casting a light on those elements not previously assessed.

With personalised (often synonymously termed as "precision") medicine at the forefront of medical research today, the question of the sources of our understanding in this field is significant. Presenting a large collection of data regarding human individuality, considerable proportions of which can be attributed to his own research, Roger J. Williams and *Biochemical Individuality* are often discussed within this context; his personal studies on the subject are therefore additionally subject to deliberation. This research aims to reconstruct and outline the origins and development of Roger J. Williams' concept of Biochemical Individuality. It attempts to outline the ways in which the nature of his research and its applications change throughout his career, seeking to characterise potentially significant modifications to his theories predicated on three leading questions.

1. In what way do Williams' ideas and his concept of Biochemical Individuality progress throughout his research and to what extent do his publications reflect this?
2. How do Williams' theories relate to the political, cultural, scientific, and historical context of the time and how does the work of fellow scientists and academics influence Williams' research and theories?
3. Does *Biochemical Individuality: The Basis for the Genetotrophic Concept*, the book most often described as Williams' definitive work on individuality,

reflect the notions of BI which predated its publication, and in what form does this final product represent a true crystallisation point in his research?

To this end, the bibliographies of Williams' previous publications will be evaluated and compared to the citations made in *Biochemical Individuality* in order to relate the importance of earlier publications to the final product. Furthermore, Williams' own publications cited in *Biochemical Individuality* will be discussed.

Due to the scope and scale of this thesis, there will be no further discussion or analysis of Williams' work following the publication of *Biochemical Individuality: The Basis for the Genetotrophic Concept* (Williams, 1956a). This book is often portrayed as Williams' primary and most important publication on BI in prior analyses and by Williams himself, with all following work deemed to be based upon its contents (Gadebusch Bondio, 2015, p. 19; Hodge and Williams, 1980a; Michl, 2015, p. 71). It is, therefore, widely seen to be the crystallisation point of Williams' research on individuality, is the work most often quoted in connection with his research of BI, and forms a natural break within the context of his career. The contents of any major publications by Williams in later years are only discussed where relevant to the above research questions. The reviews of *Biochemical Individuality* and statements regarding Williams' early work on human variation are the only documents published after 1956 considered, as they bear relevance to the reception of the work itself. Predating notions of human individuality have been subject to in-depth discussion and summarised elsewhere. These shall generally, therefore, not be deliberated further as part of this analysis.[14] Where pertinent to Williams' concepts of Biochemical Individuality, however, preceding ideas of biological variation will be discussed regarding their importance to Williams' ideas and their development. Similarly, this analysis does not aim to subsume a differentiated discussion of all contemporary theories encompassing individuality, restricting itself to the discussion of concepts pertinent to Williams' work.

---

14   For further reading, see Abrahams and Silver (2011), Gadebusch Bondio (2015), Gadebusch Bondio and Spöring (2017), Jain (2009), Jones (2013a), Michl (2015), and others.

2 Materials and Methods

2.1 Source Discussion and Research Methodology

The increasing relevance of Biochemical Individuality in medical theory and practice today marks Roger J. Williams and his theory of Biochemical Individuality as an appropriate subject for historical academic deliberation. As a well-known and widely published biochemist, Williams' resume boasts an ample and varied collection of source material for study. The structure provided by the website dedicated to his research and legacy provides a framework for source research, also making otherwise difficult-to-procure primary materials readily available.[15] Further unpublished materials, housed in the Dolph Briscoe Center for American History at The University of Texas at Austin, including one-hundred and twenty-four shelf-feet of material, additionally augment the available sources.

Reaching from fundamental biochemical research to pieces posing sociological and political theories alongside discussions of highly contested and topical subjects including, but not limited to, communism, racism, and alcoholism, Williams' publications offer substantial opportunities for consideration and analysis. His works are published in a wide variety of prominent journals and magazines, including *The Lancet*, *The Journal of the American Medical Association*, and *Vogue*. As the discoverer of multiple B-vitamins, recipient of countless awards and accolades, and holder of significant positions in scientific societies, Williams was doubtlessly a prominent figure within the scientific community (Davis et al., 2008, pp. 9–10). These honours are largely attributed to his prodigious work as a classical biochemist, though the largest portion of his publications and career are dedicated to Biochemical Individuality – a research focus for which he was not undisputed. Finally, the prominence of *Biochemical Individuality: The Basis for the Genetotrophic Concept* (Williams, 1956a) heightens the relevance of its scientific basis.

The evidence discussed within this this work encompasses all of Roger J. Williams' publications produced between 1919 and 1956. His articles and books published in this timeframe are analysed regarding their relevance to the origins and development of BI as a concept. The Biochemical Institute at the University of Texas at Austin provides a detailed list of Williams' publications

---

15 See Davis (2003a).

on its website (Davis, 2003b, 2003c, 2003d). This list forms the framework for the research for this thesis, indicating which articles were published within the selected timeframe. Reprints of the listed articles were made available by Donald R. Davis, a former colleague of Williams and the administrator of the official website dedicated to his work. Further digital research using various combinations and variations of the search terms "Biochemical Individuality", "Individuality", "Genetotrophic", "Genetotrophic Concept", "Genetotrophic Principle", "Williams", "Roger Williams", and "Roger J. Williams" in the databases SCOPUS,[16] PubMed,[17] Google Scholar,[18] BONNUS,[19] ZDB,[20] and the online archive databases of the University of Texas at Austin[21] and Oregon State University[22] produced a number of articles not included in the institute's list, as well as digital copies of the articles not available as physical reprints and secondary literature. Original editions of Williams' published (text-)books were procured via an online second-hand book platform.[23] Scientific publications by other authors are discussed when important to the development of BI, as indicated by repeated reference by Williams himself, and were also attained using the aforementioned online databases. All cited publications have been procured in their original form.[24]

Williams' published articles and books serve as the primary source of information regarding his theories. These are studied and discussed in chronological order, comparing each publication and its contents to those published previously as well as with the book *Biochemical Individuality*. Marked deviations from previous publications are subject to in-depth discussions, with certain publications highlighted due to their particularly innovative nature. Where publications signify a turning point in regard to certain ideas or theories, this is emphasised accordingly. Where bibliographical overlaps occur between publications and *Biochemical Individuality*, these overlapping citations are examined with regard to

---

16 See Elsevier B.V. (2023).
17 See National Center for Biotechnology Information (2023).
18 See Google (2023).
19 See Universitäts- und Landesbibliothek Bonn (2023).
20 See Deutsche Nationalbibliothek (2023).
21 See Briscoe Center for American History (2023).
22 See Oregon State University Libraries (2023).
23 See AbeBooks Inc. (2023).
24 All direct quotations reflect the original grammar, spelling, and emphasis of said publications. Where possible, first editions of Williams' books were used. When not accessible, the earliest possible versions were procured.

their relevance and merit. This analysis attempts to produce a profile of Williams' primary ideas of individuality, following the progression of his thoughts and theories throughout his studies from unformed ideas to concrete theories. It will then attempt to formulate a definition of Biochemical Individuality during the various stages of Williams' career, in order to summarise the information gained from these aspects of his research. The alterations to BI's definition are indicative of the extent to which Williams' concepts progress following significant publications. Wherever archive materials can provide supplementary information, this material is presented and discussed in connection with the previously described analysis.

Williams' academic estate is housed in the Dolph Briscoe Center for American History at The University of Texas at Austin.[25] Prior to the research for this thesis, the collection was largely unsighted and uncatalogued. Following a preliminary sighting by the staff of the Briscoe Center, fifty-one boxes from the "Roger Williams Papers" were selected due to their potential relevance to this dissertation. These were read, analysed, catalogued, and recorded over a one-month timeframe. Sources obtained from the archive include personal notes, correspondence, audio and video recordings, and manuscripts. Aside from the aforementioned primary sources, peer reviews showcasing the reception of Williams' published theses are also studied, as they offer an impression of the scientific zeitgeist and provide context. An interview conducted with an academic contemporary of Williams, Donald R. Davis, using the semi-structured interview technique as outlined by Harrell and Bradley (2009),[26] additionally augments the aforementioned research material and is discussed in order to provide first-hand insight into his guiding principles and working methods.

Williams' published and unpublished works have been analysed through close reading, with special focus paid to the terminology used in all documents. Changes in said terminology are examined as to whether they may serve as an indicator for changes to the material and content of his theories. The works are discussed and presented in chronological order. Marked alterations and important turning-points to the concept are noted and provide an overall structure

---

25 In January of 2023, the author spent one month in Austin for the study of these files, funded in part through the William and Madeline Welder Smith Research Travel Award of the Briscoe Center.
26 This style of interview was chosen as it encourages the development of free conversation (Harrell and Bradley, 2009). This was deemed the most useful form for obtaining an awareness of Williams' attributes as a director and colleague in conversations with the supervisor of this thesis.

along which the development of BI is examined. All numerical scientific data presented as part of these publications has been analysed as to the extent of variation shown therein, even when no mention of such statistical variation is made in the text. This serves to indicate Williams' own production of evidence pertaining to individuality. Furthermore, the bibliographies of Williams' published works are drawn upon, evaluated, and structured. This may indicate the relevance of other theories, research, and individuals for the development of BI. Where there is an overlap between the works, it is presented in the form of tables, cross-referencing and comparing all sources cited by Williams. This can show the extent of their correspondence with one another and with the final work *Biochemical Individuality*, additionally indicating the extent to which, if at all, *Biochemical Individuality* represents the cumulative result of Williams' prior research. The relative importance of Williams' own work in the overall context of BI is additionally explored. Similarly, all of Williams' publications are categorised according to their year of publication and relevance to BI. This information is presented in the form of a bar graph, to indicate the relative importance of BI within Williams' work throughout his career.

Additionally, this thesis endeavours to discuss Williams' work within the sociological and scientific timeframes of the pre-, intra- and post-World War Two eras alongside their corresponding effects on his work and vice versa. Where pertinent, important theories on which Williams bases his experiments or theories are explored further. Their influence is studied within the context of BI, with differences and similarities between these and Williams' works illustrated. This thesis additionally attempts to discern the influence of other academic theories and contemporaries on Williams' concepts by investigating the extent to which certain individuals and their research may have had an effect on his own thoughts. With a wide variety of peer-reviews available for large parts of Williams' career, the reception of his work forms a further pillar of research, with the influence of such external critique and praise explored.

All aspects of BI, including the potential importance and influence of other persons and non-scientific ramifications of the theories presented, are assessed and analysed in regard to differences and similarities to the book *Biochemical Individuality: The Basis for the Genetotrophic Concept*. This book is hypothesised to be the crystallisation point of Williams' research on BI, in which all previously collected information is presented. Where appropriate, the historical, cultural, political, and scientific context of the documents in question is provided, with aspects of Williams' work pertaining to these modifying circumstances explored accordingly. Additionally, the reception and reviews of Williams' publications, and possible consequent changes made to the paradigm based thereon, are discussed to gauge the novelty of Williams' concepts within the scientific zeitgeist. To accurately depict alterations to the concept of BI over time, a brief definition of Biochemical Individuality is provided for the respective timeframes at the end of each chapter.

3 A Career in Biochemistry

The object of this dissertation is to systematically depict the origins and development of Roger J. Williams' most defining and widely published theory: the concept of an innate Biochemical Individuality existing in all human beings, generating variation between individuals with far-reaching consequences for science and society. While Roger Williams' publications themselves provide evidence towards the genesis and progress of the concept, a thorough examination and contextualisation providing valuable insight into the origin and development of Biochemical Individuality does not yet exist. The object of the following dissertation is neither to portray a curriculum vitae of Roger J. Williams, nor to document his life and work in complete chronological order. Therefore, the impressions supplied by a biographical piece provides a framework within which the discussed research and publications are to be reflected. This chapter shortly outlines Roger Williams' life and career, providing an overview in which his research on Biochemical Individuality is to be read.

3.1 Upbringing and Education in Biochemistry

Born as sixth child of missionary parents in India on the 14th of August 1893, Roger J. Williams' upbringing and consequent adulthood are characterised by frequent relocations (Davis et al., 2008, p. 4). Growing up in rural California and Kansas, his adolescent years are dominated by an exploration of nature and moderate scholastic success, due to an affliction of the eyes called "aniseikonia" (Williams, 1954a).[27] Following studies at the University of Redlands, Williams receives his Bachelor's degree in 1914, swiftly followed by a high school teacher's certificate from the University of California, Berkeley in 1915 (Davis et al., 2008, p. 5). Working as a high school teacher subsequent to a short stint as the foreman of a small nursery, Williams spends the two ensuing years in Hollister, California, before joining the University of Chicago and graduating with a Master's degree in 1918 (Davis et al., 2008, p. 5). His Ph.D. thesis entitled "The Vitamine Requirement of Yeast" is published in 1919 (Williams, 1919).[28]

While the development and history of biochemistry has been outlined elsewhere on multiple occasions,[29] the state of the subject at the time of Williams'

---

27 See Section 3.2. Academic Networks.
28 See Section 4.2. "The Vitamine Requirement of Yeast".
29 For further information, see Kohler (1975).

studies bares relevance to his later career. With biochemistry arising as a self-conscious speciality in the early 20th century, Williams initial scientific education falls into an era of fundamental discussions within the research field as a whole (Kohler, 1975, p. 290). Contributing to a phase of great discovery and frequent revolutions within the field, the necessity and new-found possibilities for the development of new research techniques play an important role in Williams' biochemical discoveries.[30] Working within an era in which biochemical research focuses on enzymes and metabolic pathways, Williams' career sees the emergence of a vast amount of biochemical data and research (Kohler, 1975, p. 294). This greatly contributes to Williams' efforts on Biochemical Individuality, as the analysis of his contemporary research provides much of the data he uses as evidence.

Parallel to these biochemical strides, the development of pharmacogenetics and human genetics as scientific subjects additionally influence Williams' work. Following from Archibald Garrod's deliberations on "inborn errors of metabolism" and Beadle and Tatum's "one gene – one enzyme" hypothesis,[31] Williams' research coexists with, and in part influences, the development of these new research fields (Motulsky, 2002, p. 684). Though his work is mentioned by multiple important personalities of the field, Williams' research must be seen as separate due to his deviating research focus (Kalow, 1962; Motulsky, 2002, 2010; Vogel, 1959). While recognising heredity as an important factor in metabolism and disease, Williams' research does not attempt to ascertain exactly which genetic factors have an influence on his studies. With the field of genetics also fundamental to Williams' Genetotrophic Concept,[32] the importance of these parallel developments is legion.[33]

Joining the University of Oregon as an assistant professor in 1920, Williams begins his academic career with the principal biochemical study of vitamins (Fowler, 1988, sec. D, p. 30).[34] Becoming a professor in 1928, Williams continues his career at Oregon State College from 1932 to 1939, largely concentrating on the structure of B vitamins and announcing his discovery of pantothenic acid in 1933 (Davis et al., 2008, p. 6).[35] Arriving at The University of Texas at Austin in

---

30 See Chapters 4. Origins of Biochemical Individuality as a Concept, 5. Vitamin Studies, and 9. Human Individuality.
31 See Section 6.4. "Distinctive Metabolic Traits".
32 See Chapter 8. Genetotrophic Disease.
33 For further information, see Motulsky (2002, 2010).
34 See Chapter 5. Vitamin Studies.
35 See Chapters 4. Origins of Biochemical Individuality as a Concept and 5. Vitamin Studies.

1939, Williams primarily accepts the position of professor in the Department of Chemistry, before founding the Biochemical Institute in the following year (Biesele, 1988, p. 53). Serving as the director of said institute for a total of 23 years, Williams oversees the discovery of more vitamins than any other biochemical laboratory in this timeframe (Davis, 1988, p. 123). Best known for his early discovery of pantothenic acid, the largest portion of Williams' studies focus on the intricacies of human individuality on a biochemical level. Remaining academically active until his death on the 20th of February, 1988, Williams publishes a total of approximately 300 academic articles and 21 books throughout his career (Davis et al., 2008, p. 3).

3.2 Academic Networks

While the following chapters will discuss individual theories which influence the development of Williams' concept of Biochemical Individuality, the role of meaningful individuals and their influence on Williams' work is a similarly important aspect to his career. One defining characteristic of Roger Williams' scientific discoveries is his tendency to rethink and reform, not shying away from new ground and uncommon techniques. His major breakthroughs are reached by avoiding convention and devising new methods of research. Williams regularly relies on the advice and opinions of multiple colleagues and co-workers during these pioneering studies. Often sending manuscripts and pre-prints to those whose opinions he values, changes of scientific, linguistic, and organisational nature can frequently be observed following their suggestions in Williams' writing process. The influence of these individuals cannot be gauged according to uniform criteria, as their personal, academic, and professional affiliations modulate their potential authority and impact. The impressions made by their comments are therefore considered within this context. The Roger Williams Papers at the Dolph Briscoe Center for American History contain the entirety of Williams' correspondence from throughout his career; here certain individuals who are regularly contacted for advice and offer influential opinions stand out.

An inescapable factor in Williams' career is that no matter how significant an individual and their advice might seem, a physical aggrievement will always be a limiting factor for their influence. Suffering from aniseikonia, prolonged reading presents a challenge to Williams from a young age.[36] Describing the act

---

36 Aniseikonia is an affliction oft the eyes, in which perception of the size and shape of objects is inhomogeneous. This often makes reading very strenuous for affected patients. For further information, see Schmidt-Erfurth and Kohnen (2018).

of reading as "like walking uphill dragging a log" (Williams, 1954a), reading for longer than 15 minutes is repeatedly described as being extremely strenuous to Williams (Williams, 1950a; Hodge and Williams, 1980b). His pioneering position in later years is often attributed to this lack of extensive reading, as his work is less tainted by the opinions of the scientific community (Davis, 1988; Williams, 1954b, pp. 66–67). With limited influence by the scientific zeitgeist, Williams seeks out personal definitions and answers rather than relying purely on the work of others (Williams, 1954b, pp. 66–67). He himself claims to promote and accentuate thinking in his work, preferring to think, rather than read, things through (Hodge and Williams, 1980b). Visiting libraries only to answer specific questions he had already thought about extensively, he describes often being disappointed by the lack of answers these visits provide (Hodge and Williams, 1980b). Even following multiple correctional operations and interventions, his eyesight continues to be an issue due to him developing macular degeneration, preventing him from becoming an avid reader (Hodge and Williams, 1980b).

The importance of Williams' widespread academic network therefore bears relevance in light of his distinctive approach to the scientific method. With extensive reading rendered an impossibility, Williams' various correspondences often contain reflections on his work and suggestions for structural alterations, as well as discussions on the merit of certain evidence used. His correspondence within a network of academic contemporaries takes place in the form of shorter, less optically strenuous, deliberations. The influence of others on Williams' work must, therefore, always be seen within this context.

While considering the aforementioned limitations and influences on the effect of others on Williams' research, multiple distinctive individuals project themselves through their correspondence and critique. As Director of the Division of Natural Science and Agriculture at The Rockefeller Foundation, Warren Weaver (1894–1978) is one of the names found most often in Williams' correspondence.[37] The topics discussed by the two men range from organisational aspects of the funds granted to DBUT by The Rockefeller Foundation to specific discussions of manuscripts and pre-prints sent to New York City for Weaver's perusal.[38] Weaver, a mathematician by training (Rees, 1987), is notably much more critical

---

37   While trained as a mathematician, Weaver later worked for The Rockefeller Foundation and closely with Williams regarding his grant from the aforementioned organisation. For further information, see Rees (1987).
38   The Rockefeller Foundation granted The Department of Biochemistry at The University of Texas at Austin 25000$ for the research of biochemical aspects to alcoholism in 1949 (Rhind, 1949).

of Williams' works than other colleagues. Openly commenting on less favourable aspects of his papers and manuscripts, Weaver's annotations are received as highly valuable and often lead to changes in the final products (Weaver, 1945, 1946; Williams, 1953b, 1954c). Exemplary for this influence is Weaver's role in the changing of the term "humanology" to "humanics" in *The Human Frontier* (Weaver, 1945, 1946; Williams, 1946b).[39] As a regular contributor and often contacted advisor, Weaver's constructive critique has a positive effect on Williams' work, often reigning in his more extreme works and theories (Weaver, 1954). Causing Williams to rethink and justify his opinions and theories, Weaver is a source of reflection (Williams, 1954c). The funding Weaver's division supplied is equally significant, boosting Williams' stock and providing the financial support necessary for his research (Hodge and Williams, 1980b). He is, therefore, to be seen as one of the most influential individuals for Roger Williams' work.

---

39   See Section 6.2. Audience.

THE ROCKEFELLER FOUNDATION
49 WEST 49th STREET, NEW YORK 20

January 28, 1946

THE NATURAL SCIENCES
WARREN WEAVER, DIRECTOR
FRANK BLAIR HANSON, ASSOCIATE DIRECTOR
HARRY M. MILLER, JR., ASSISTANT DIRECTOR

Dear Dr. Williams:

    I have read your revised introductory statement and first chapter. I am returning it to you herewith with various comments. I have not pulled punches at all in making these comments - on the assumption that it is friendly and helpful to make these comments directly to you and in advance of publication. I recognize, of course, that my own viewpoint is both prejudiced and limited.

    About "humanology". It sounds to me too much like a word made up by an advertising man. One very able person here who read part of your manuscript said: "Perhaps I am just misled by my dislike of trying to solve problems by rubric phrases such as humanology." The phrases "the science of man", or "the science of individuals" (if you want to emphasize the individual as contrasted with man-in-the-abstract) seem to me dignified and accurate.

    I do think that you have substantially improved this first part, but I continue to think that you weaken your case by taking in too much territory, claiming too much, and condemning too much. And I begin to be a little specially confused as to what your purpose is in writing this book. Perhaps my criticisms are wholly inappropriate because I am misjudging purpose. The manuscript is beginning to sound more and more as though it were written primarily (and almost exclusively) for the purpose of interesting and arousing the general public. I keep thinking of the manuscript in terms of its effect on and reception by scientists. If it is written exclusively or primarily for the general public, and you do not care particularly what the scientists think, then my comments largely collapse. But if you are writing for the public, then it is fair to question what will be accomplished. Perhaps a general stimulation of public interest would have important reverberations? Perhaps Henry Ford would decide to spend his money this way? Perhaps the public would demand

Fig. 1: Letter by Warren Weaver to Williams on 28.01.1946 critiquing Williams' use of the term "humanology" in an early version of The Human Frontier, first page; Dolph Briscoe Center for American History, The University of Texas at Austin, Roger Williams Papers, Box 88-087/26a, Folder: Correspondence Concerning Humanics Sept. 1945 – March 1946

2.

    increased emphasis by universities, etc.? I just don't know. A campaign to get something started, however, is a little difficult to wage with the general public.

    This has been, to this point, a personal letter. By way of institutional comment, I would say that I am reasonably sure that The Rockefeller Foundation would not consider becoming an operating agency in this field. We wholly sympathize with your purpose, and admire your drive and interest. But I do not think we would be prepared, at least in any near future, to discuss support of an institute of the sort you envisage.

                      Very cordially,

                      Warren Weaver.

Dr. Roger J. Williams
Director, Biochemical Institute
University of Texas
Department of Texas
Austin 12, Texas
ww.ahc

Fig. 2: Letter by Warren Weaver to Williams on 28.01.1946 critiquing Williams' use of the term "humanology" in an early version of *The Human Frontier*, second page; Dolph Briscoe Center for American History, The University of Texas at Austin, Roger Williams Papers, Box 88-087/26a, Folder: Correspondence Concerning Humanics Sept. 1945 – March 1946

Robert R. Williams (1886–1965), the older brother of Roger J. Williams, is a further name often encountered within the Roger Williams Papers.[40] A famous

---

40   Robert R. Williams was a biochemist and is most famous for his research surrounding Beri Beri. For further information, see Sigma Xi: The Scientific Research Honor Society (2023).

chemist in his own right, the two siblings are in regular contact regarding all of each other's publications as well as personal matters. The two brothers exhibit a friendly relationship, offering honest opinions and suggestions for the improvement of their respective works. Serving as an inspiration to Roger Williams in his choice of vocation, Robert Williams is a constant source of advice for his younger brother (Williams, 1954a). As an older brother who "feel[s] free to criticize" (Williams, 1954c), endorsements from Robert Williams are taken as a stamp of quality and confirmation of validity for theories and papers. He therefore may not be omitted from a list of influential individuals. The entirety of their correspondence can be found in the Roger Williams Papers, Box 88-087/8, Folder: "Williams, R. R.".

A "most dominant" individual at the start of Williams' career is chemist Julius Stieglitz (1867–1937) (Williams, 1954a).[41] As his first mentor and president of the American Chemical Society (Hodge and Williams, 1980b), Stieglitz has a pronounced effect on Williams' understanding of organic chemistry, later inspiring Williams' first successful textbook (Williams, 1954a).[42] Additionally, Williams derives his teaching methods from the observation of Stieglitz' classes, albeit altering certain aspects to better suit his own preferences (Hodge and Williams, 1980b). Stieglitz is an important figure in Williams' chemical work, though his effect on the development of BI is minor due to his passing early in Williams' career. Similarly, biochemist Fred Conrad Koch (1876–1948) has a pronounced effect on Williams' early career.[43] Offering him his first fellowship at the University of Chicago and thereby allowing him to complete his thesis work, Koch is involved in Williams' first well-known publication and influences his later decision to work in biochemistry (Williams, 1954a).

Benjamin Clayton (1882–1978) is one of the most important figures throughout Williams' career, though his contributions are not academic in nature.[44] His foundation provides Williams with the first meaningful funding for the

---

41 Julius Stieglitz was a chemistry professor and president of the American Chemical Society in 1917. For further information, see Noyes (1939).
42 See Section 4.4. Yeast Extract Research.
43 Fred Conrad Koch was a chemistry professor at the University of Chicago and the editor of *Archives of Biochemistry*. For further information, see Hanke (1948).
44 Benjamin Clayton established the Clayton Foundation for Research, now known as Clayton Biotechnologies, Inc., in 1933 (Clayton Biotechnologies, 2023). The Foundation offered financial support to the Biochemical Institute at the University of Austin throughout Williams' career as of the 13th of September, 1940 (Williams et al., 1966, p. 2).

establishment of DBUT (Hodge and Williams, 1980b; Williams et al., 1966, p. 2). This funding is essential to the development of BI, as it enables Williams and his colleagues to study topics of their choosing (Williams et al., 1966, p. 2). Primarily interested in the biochemical basis of cancer, Clayton does not restrict his funding to DBUT, which was later named "The Clayton Foundation Biochemical Institute" in his honour, to working on only this subject. *Biochemical Individuality* is dedicated to Benjamin Clayton, describing his moral and material support as invaluable (Williams, 1956a, p. vi).

Linus Pauling (1901–1994), noted chemist and Nobel-laureate, is a further regular correspondent throughout Williams' career.[45] He offers feedback and advice on Williams' books *The Human Frontier* and *Biochemical Individuality* and is regularly sent manuscripts by Williams (Pauling, 1946, 1957). Similarly, both scientists ask for each other's advice on technical matters (Pauling, 1937; Williams, 1936). When Williams is disenchanted with his work in Oregon, Pauling acts as a personal friend and confidant.[46] Their connection continuing throughout their lifetimes, the two scientists work together closely later in their careers (Oregon State University Libraries Special Collections & Archives Research Center, 2014a, 2014b).[47]

While the previous paragraphs have depicted the role of individuals from outside of Williams' own institute at the University of Texas, the role of his colleagues at DBUT must additionally be appreciated. Williams' publications with co-workers are legion, though the influence of the respective co-authors of these articles and books is impossible to gauge from the content of the documents alone. The lack of research allocation at DBUT, and the importance of his co-workers' academic freedom to Williams, are discussed elsewhere.[48] Former colleague Donald R. Davis describes how this influences his input on individuality from within his own biochemical institute:

---

45 Linus Pauling was famous chemist and recipient of two Nobel prizes; one for chemistry and the other for peace. For further information on the life and work of Linus Pauling, see Dunitz (1997).
46 See Section 5.2. The Vitamin Content of Tissues.
47 The correspondence and collaboration between Roger Williams and Linus Pauling have been summarised and discussed in blog form by the Oregon State University Libraries Special Collections & Archives Research Center. For further information, see Oregon State University Libraries Special Collections & Archives Research Center (2014a, 2014b).
48 See Chapter 5. Vitamin Studies.

> As far as I know, none of his colleagues were very much focused on individuality. I know that he did share manuscripts with many colleagues, asking them to review and comment on them. But I don't think he had any "soulmates" so to speak, to talk about individuality. He had a long-term office assistant, Margaret Biesele; she probably was a sounding board for him.[49]

The influence of Williams' colleagues is, it follows, largely reduced to their comments on manuscripts. Their additional role in devising and performing Williams' scientific research is indicated elsewhere.[50] Though Williams' pioneering role in regard to his concept of BI causes the possibilities for collaboration and discussion to be minimal at best, the value of these collegial reviews, so Davis, is nonetheless great, as suggestions are often integrated into later versions of Williams' publications and amplify the longevity of his written works.[51]

Though Williams' ophthalmologic affliction may indicate a certain independence of thought regarding his conceptualisation of BI, his widespread personal correspondence shows the role of others in the development of said theory. A range of scientific colleagues and confidants provide valued feedback on all of Williams' major publications, often leading to modifications of his work. The changes to the term "humanology" exemplify the influence of collegial suggestions. Outside of the academic realm, Williams' longstanding financial supporter provides a basis on which his work can safely rest. These contributions do not, however, pertain to the central themes of Williams' works or the concepts presented therein. Influences of prior theories aside,[52] Williams' concepts and ideas are always fundamentally his own, though others play a role in augmenting these.

---

49 See Section 16.1. Interview with Donald R. Davis.
50 See Section 8.2. Collaborative Individuality.
51 See Section 16.1. Interview with Donald R. Davis.
52 See Section 6.4. "Distinctive Metabolic Traits".

4 Origins of Biochemical Individuality as a Concept

4.1 Youthful Observations

> My attention was first attracted to these individual differences, and particularly the sense of touch, when I was about four years old. I was with my father in an orchard where ripe peaches were abundant, when he noted that although I liked the taste of peaches I could hardly be induced to touch one. The fuzzy skin made me cringe. My father was quite indifferent, himself, to this fuzz, but he told me that I had probably inherited my dislike from my paternal grandfather who had exactly my reaction to the skin of peaches. (Williams, 1946a, p. 77)

Reminiscing in his book *The Human Frontier*, Williams describes his earliest conscious realisation of the differences which define us as individual human beings. The simplistic description of differences is fitting, as it is not in the nature of most four-year-olds to think in a particularly scientific manner. Though this portrayal is, admittedly, of anecdotal importance only, it is an indication of the longevity of Williams' own development of the idea.

In addition to anecdotal evidence of first thoughts, this excerpt contains a second contextually important factor. The layman-like description of the inheritance of sensory traits by Williams' father is archetypal for the attitude towards individual likes and dislikes of the time. In 1897, the work of Gregor Mendel (1822–1884) was largely unknown in scientific realms, let alone to the wider public (Mendel, 1866).[53] Williams' father's premonitions on the inheritance of human traits were therefore, at the time, of a merely intuitional nature. These intuitions are of essential importance to Williams' process as they, while not scientifically proven at the time, are the presumptions on which his primary thoughts on the subject are based.

---

53 First published in 1866, Mendel's work on fundamental genetics lay forgotten for more than forty years, when the laws it proposed were rediscovered around the turn of the century by three separate research teams (Gayon, 2016). For a short summary, see Gayon (2016). For more extensive works, see Rheinberger and Müller-Wille (2009) or Jacob (1993).

4.2 "The Vitamine Requirement of Yeast"[54]

Throughout his postgraduate research, Williams lays the foundation for the work that would later define his biochemical legacy. Publishing his first ever manuscript in 1919, Williams begins his academic career with the study of fundamental biochemical principles. While studying the growth patterns of yeast in various environments, the fact that some yeast cells grow faster than others in an identical milieu becomes apparent (Williams, 1919). The behaviour of these cultures, though often alike in looks, provides a visual correlation for the variation between individual cells (Williams, 1954b, p. 66). Williams' research at said point, however, focuses on the effects of the cells' milieu rather than on the cellular components themselves. This justifies his lack of further comment on these growth patterns in his dissertation.

In retrospect, Williams voices the suspicion that the observations of this early work acted as a forerunner to his later interest in BI (Williams, 1954b, p. 66). He himself states that these first observations "later caused me to suspect that every living creature – human, animal or plant – is anatomically and biochemically unique" in a later interview (Lampe, 1982, p. 22).[55] This statement confirms the idea that BI and its possible implications, not a focus of research in the early 20th century, only later appear noteworthy enough to solicit further thought. It is likely that Williams did not realise the discrepancy in cellular growth patterns could be related to their individual metabolisms at the time. In hindsight, the results of this research offered relevance and evidence of the budding concept of BI.

Williams' work on his thesis is not only noteworthy due to the relevance later attributed to its data. The method of biochemical research presented in "The Vitamine Requirement of Yeast", utilising the study of microorganisms instead of more complex animals, was uncommon at the time. In his description of the history of The Department of Biochemistry at The University of Texas at Austin, Williams deliberates on the feedback he received for his work with yeast.

---

54 (Williams, 1919) To improve the clarity of this section-title and avoid confusion, the citation for this quote is presented in this footnote, deviating from the citation style used in the rest of this dissertation. The full title of this article is "The Vitamine Requirement of Yeast: A Simple Biological Test for Vitamine".

55 In the published interview, this quote is prefaced by a statement claiming that the observations it comments upon occurred 65 years prior (Lampe, 1982, p. 22). This pinpoints these observations to 1917, the year in which Williams' research for "The Vitamine Requirement of Yeast" was his scientific focus.

Describing advice received from "a very prominent investigator" who remains unnamed, Williams (1954b, p. 67; 1966, p. 4) is urged to continue his studies with more complex species instead. This would have been the "settled" methodology for research of this kind at the time (Williams et al., 1966, p. 4). Even in the early stages of his career, he appears unafraid to break with traditional scientific procedures.[56] Working with complex organisms for basic biochemical research would be unthinkable in contemporary research, while the approach taken by Williams is the unusual one at the time (Williams et al., 1966, p. 4). Williams (1941a, 1944) himself later publishes articles on the extensive importance and relevance of the use of microorganisms in vitamin research.[57] This willingness to experiment with and develop new methods for research rather than apply established techniques is a defining aspect of Williams' work. Furthermore, it is an early indicator for his willingness to propose new and controversial ideas rather than merely continuing and expanding on the work of others.

The search for ideal nutrition becomes a central field of study for Williams in later life. Applying his previous search for the ideal nutrition of yeast, optimal human nutritional organisation and supplementation becomes a topic of profound interest (Hodge and Williams, 1980b).[58] The insights stemming from

---

56 Though his work using microorganisms for his research on vitamins was of pioneering character, Williams was not the first to apply this method. Williams (1941a, p. 413) describes the first use of such microorganisms within the realm of vitamin research in the work leading to the discovery of "bios" by Manille Ide. Williams remained convinced of the pertinence of this method. Evidence thereof is provided by his own vitamin research, which is largely based upon work with yeast. In an article written in cooperation with two paediatricians and two biochemical colleagues, Williams describes how microbial experimentation made vitamin analysis "easier and less time consuming", in comparison to "laborious animal assays" (György et al., 1941, p. 477). This statement reappears in a later symposium-discussion (Williams, 1944, p. 135).
57 Williams describes the importance of microorganisms for both the previous and future research of vitamins. Emphasising the futility of animal-based research in the biochemistry of vitamins, he makes the case for further use of microorganisms, suggesting new and altered techniques could provide a potent basis for scientific discovery (Williams, 1941a). Furthermore, the statement, "that microorganisms will play an important part in the discovery of these [vitamins] seems evident" underlines the significance of this research approach (Williams, 1944, p. 126).
58 Such interest is indicated by a plethora of publications upon the aforementioned topic of nutritional supplementation, such as *What To Do About Vitamins* (Williams, 1945a) and *Nutrition and Alcoholism* (Williams, 1951a). For a full lists of publications, see Davis (2003b, 2003c, 2003d).

his first body of research can be regarded as Williams' first inkling of an idea of individuality in a biochemical sense. His previous, adolescent observations are merely based around the individuality of his own experience (Williams, 1946a, p. 77). This first scientific association with individuality later produced a much more profound understanding. The first consideration of uniqueness on a cellular level may not have been as fully developed as those ideas which followed, the historical origin of the concept of BI, however, is clearly characterised by increased knowledge and complexity achieved through primarily random discoveries. Having noticed a basic propensity for individuality among humans in youth, Williams' additional reflexions on by-the-by observations in his postgraduate experimentation can be considered a vital steppingstone on his path to a concept of BI.

4.3 Adverse Drug Reaction

As is often the case in scientific research, there is an aspect of coincidence to Williams' increasing interest in BI. Happenstance leads to further interest in a field which may otherwise have gone untouched. In the early 1920s, Williams undergoes an abdominal ulcer surgery.[59] While awakening from the anaesthetic, he is administered morphine to alleviate his pain. This drug, which usually has a pain-relieving and sleep-inducing effect on "normal" individuals, seems to show an adverse effect when confronted with Roger Williams' metabolism.[60] While his pain is alleviated, Williams' retrospective of the experience is described in an interview from 1982:

---

59   The exact year in which this operation took place is disputed. Multiple articles and primary sources cite various years, though all of these are in the early 1920s. David Lampe's (1982) article places the operation in 1920, while (Davis, 1988) places the ulcer operation in "about 1921". Williams (1977, p. 57) describes the timing of the operation vaguely in *The Wonderful World Within You: Your Inner Nutritional Environment*. Relating this event to the publication of Barry Anson's (1950) *An Atlas of Human Anatomy*, he claims it was published "more than twenty years after the morphine episode" (Williams, 1977). This also places the operation and following experiences somewhere in the 1920s, though an exact date cannot be ascertained at the time of writing.

60   The term "normal", while inaccurate, is used explicitly here, as it reflects the opinions of the time. As Williams tends to criticise in his later career, most doctors and society in general considered there to be a dichotomy of "normal" and "abnormal". Williams' conclusion that "all of them are in a sense 'abnormal'" diverges from this view (Williams, 1956a, p. 3).

> My mind became so active that it was racing from one thought to the next. I was then given a larger dose of morphine. All night long I was suffering continual mental torture. Why did I react this way to morphine? The doctor assured me it was merely an idiosyncrasy. Nothing in the library could give me a clue [...] The experience aroused my scientific curiosity. There must be a reason for my reaction. (Lampe, 1982, p. 23)

This event has an encouraging effect regarding Williams' interest in BI. Idiosyncratic, defined as "specific to an individual", is a popular explanation for unsolved medical mysteries (Uetrecht and Naisbitt, 2013). Williams is not satisfied with the absence of answers science can provide him with at the time. The first concept of an explanation for idiosyncratic drug reactions was formed by Karl Landsteiner (1868-1943) through his Hapten Hypothesis in 1935, 15 years after Williams noted a lack thereof (Uetrecht, 2008). Mention of "Metabolic Idiosyncrasy" as an explanation for the toxic effect of drugs was first given by Hans Zimmerman in 1976, a full 20 years after Williams had published his concept, proving how far ahead of its time Williams' concept really is (Uetrecht, 2008).[61] [62]

Experiencing what it means to deviate from the norm first-hand draws Williams' attention to the existence of human variation. In an interview and a later publication, *The Wonderful World Within You: Your Inner Nutritional Environment*, Williams (1977, p. 56; 1980a) reflects that his "keen interest in individuality began following a surgical operation", further underlining the importance of the incident. His early manifestations of an idea of individuality become substantially more acute when personally confronted with the consequences of an unexpected individual reaction. This interest in individual drug reactions is evident long before the first pharmacogenetic strides are made by Werner Kalow in the 1950s (Jain, 2009, pp. 4-5). Williams does not, however, immediately begin work on his own concept of an explanation; other research takes priority.[63]

---

61   "Metabolic Idiosyncrasy" is also referred to as "MI" within this thesis.
62   While Zimmerman does not quote Williams directly, indicating his thought process may have occurred independently from Williams' findings, his model is nevertheless heavily based on the principles of BI, demonstrating the progressive nature of Williams' work (Zimmerman, 1976).
63   This is indicated by Williams' publications following this operation (Davis, 2003b). Additionally, further textbooks and projects published in the following years provide evidence for deviating prioritisation. For a full lists of publications, see Davis (2003b, 2003e).

An adverse reaction to the administration of morphine further increases Williams' interest in interindividual differences. Though this does not immediately translate into concrete research in the form of medical trials, the experience promotes curiosity in the subject alongside other research at the beginning of Williams' career.

4.4 Yeast Extract Research

While Williams' research with a primary focus on BI first appears in the 1940s, prior work implies an on-going thought process following his medical reaction in 1920. Though earlier descriptions indicate a personal interest in BI, Williams' early published works do not reflect this interest in professional terms. His first Book, *An Introduction to Organic Chemistry*, published in 1927 by the D. van Nostrand Company Inc., was met with enthusiasm and adopted by more than 300 Universities in its first year, yet makes no allusions to individuality (Williams, 1935a, 1954a).[64] A dislike of the other available textbooks on chemistry prompts Williams to create his own according to his own structure of revision and learning, as an aid for his own efforts in teaching (Hodge and Williams, 1980b). Though the chemical structures of hormones, such as thyroxine and oestrogen, vitamins, and other organic compounds (later an important feature of his work on BI), are all discussed in a chemical sense, no notion of an individuality with relation to these is mentioned (Williams, 1935b, 1935c). Textbooks published in later years include such acknowledgements of BI.[65]

Williams' scientific focus in the 1930s lies on the study of the structure of pantothenic acid. Having accepted his first post as junior professor at the University of Oregon in 1920, Williams is not in the position to suggest ground-breaking research, especially when such a promising field of research was available to him (Hodge and Williams, 1980a). Further work and discoveries were necessary to develop the hypothesis that BI should be a field of in-depth scientific research.

---

64   The D. van Nostrand Company Inc. was one of the United States' most prominent scientific and technical publishing houses in the 20th century (Watson, 1949). Exclusively publishing books on the sciences and engineering, all of Williams' textbooks on biochemistry, excluding *The Biochemistry of the B Vitamins* (Williams et al., 1950e), are published here. The most prominent book published by the D. van Nostrand Company Inc., *Van Nostrand's Scientific Encyclopedia* (Considine, 2005), is now published by John Wiley & Sons, Inc., who purchased the company in 1997 (New Netherland Institute, 2023).

65   See Section 7.4. An Introduction to Biochemistry, Second Edition.

Though not principally designed to research such matters, Williams' continued efforts in identifying pantothenic acid, its properties, and structure would lead to discoveries indicating variability in the metabolism of organisms. Williams dedicates multiple publications to presenting new testing methods developed for his research on pantothenic acid (Adams and Williams, 1921; Williams, 1920; Williams et al., 1927). While offering new techniques for the quantification of "bios",[66] Williams comments on the incongruities between his results and those of other scientists, despite using a product seemingly identical to that used by himself and his colleagues (Williams et al., 1927):

> We could explain the discrepancy between their results and ours only on the basis that we worked with different strains of yeast. (Experiments reported in Section III below seem to bear out the reasonableness of this explanation. Different strains of *Saccharomyces cerevisiae* seem to give diametrically opposite results). (Williams et al., 1927, p. 228)

Considering the importance of reproducibility for scientific research, subsequent investigations of this possible source of error provide additional evidence for the differences between the strains of yeast. Noting these discrepancies, Williams et al. (1927, pp. 228–232) discuss the benefits of one particular strain and medium, and indicate possible procedures for future tests. Williams is not satisfied with merely providing the solution for avoiding the aforementioned discrepancies, as indicated by a follow-up article published two years later in which results of research ensuing from this work are published in the same journal.[67]

> The comparison of the behaviour of six distinctive strains of *Saccharomyces cerevisiae* toward various preparations was the result of two motivational pressures:

---

66 "Bios" is the term for a water-soluble yeast growth factor given by E. Wildiers in 1901 (Ainsworth, 1976, p. 111; Wildiers, 1901). Williams and Beerstecher (1931, p. 205) include a discussion of the term "bios" in *An Introduction to Biochemistry*, in which the original research upon as well as research based upon this discovery are described. The term itself is described as "pretentious", indicating certain contempt for the use of such a grand sounding term. This contempt may additionally stem from Williams' assertion that Wildiers (1901) was not the true discoverer of "bios", but that this feat should be attributed to his overseeing professor Manille Ide (Williams, 1938). He further describes Wildier as "immature" and "unproductive", additionally supporting the hypothesis that this critique is more of the author himself than of the scientific aspects. For further information on the disputes around and discovery of "bios", see Ainsworth (1976).

67 The article and research were completed and received by the Journal of the American Chemical Society on the 26.11.1928, however only appeared in the publication in September of 1929 (Williams et al., 1929, p. 2764).

(…) we were interested to know how important the substance in question might be; whether (…) it functions for one strain of yeast only or whether it might not be important in the nutrition of several strains of yeast. A second reason for making this study (…) was that few controlled experiments have been reported in which different strains of yeast have been compared directly with each other and we deemed it important to know just how wide a variation might be expected. (Williams et al., 1929, pp. 2764–2765)

Studies explicitly focusing on accumulating and comparing data of individual behaviours and values, rather than the average behaviour of a heterogenous group, are a defining characteristic of Williams' later work within the realm of BI. In 1928, only one example of such a research project can be found in Williams' repertoire, though it is not entirely dedicated to this research.[68] His interest in comparative research on individual yeast strains is mirrored by the importance later attributed to the comparative research of individual human beings.[69]

In the case of "The Effect of Various Preparations on the Growth of Bakers' and Brewers' Yeasts", the conclusions drawn from the study are equally as relevant as the mode of research itself, as they indicate a first understanding of BI (Williams et al., 1929). In the fourth point of their summary, Williams et al. (1929) pose the idea of metabolic differences between strains of yeast:

In spite of the uniformities above noted, each of several different strains of yeast (so-called *Saccharomyces cerevisiae*) reacts more or less distinctively toward different "bios" preparations. In some cases the contrast in behavior is very marked. This indicates possible deep-seated differences in the metabolic processes in different strains. (Williams et al., 1929, p. 2773)

This analysis constitutes Williams' first public suggestion that differences between genetically heterogenous groups could stem from metabolic differences. Though the evidence previously provided offers suggestions of interest and relevant thought processes, this article poses the first true theory of a BI between yeast factions. These nutritional discrepancies are additionally referred to in multiple articles published in the following years (Williams et al., 1933, pp. 2912–2913; Williams and Bradway, 1931, p. 783; Williams and Honn, 1932, p. 629). The proposal, however, differs from BI in one very important aspect: it does not imply a biochemical uniqueness of the individual cell, but rather of the entire strain of yeast. It can, therefore, be described as a precursor or intermediary with relation to BI. It does not, however, constitute a true first concept of BI.

---

68 See Section 4.5. "'Taste Deficiency' for Creatine".
69 See Section 6.1. "Humanity Must Understand Itself".

An additional publication of interest is "The Use of Fractional Electrolysis in the Fractionation of the 'Bios' of Wildiers", published in November of 1931 (Williams and Truesdail, 1931). Though the publication places a principle focus on the separation and fractionation of bios "into two distinct factors", the discussion of the results brought forth by this new method brings with it further relevant aspects (Williams and Truesdail, 1931, p. 4171). While evaluating the importance of multiple nutrilites for the growth of single-celled organisms, Williams and Truesdail (1931, p. 4180) extrapolate their findings and suggest an additional importance for "numerous forms of birds and mammals which are known to carry on a varied metabolism". This observation signifies Williams' primary suggestion of metabolic variation between complex organisms. As Williams includes human beings in this group on following occasions,[70] this excerpt consequently represents Williams' first mention of metabolic variation in humans. No reference to a possible source for this information is provided, it is therefore not possible to ascertain from where the assumption stems. Following this research, Williams re-evaluates his prior work leading to the first true publication of the basic concept of BI.

4.5 " 'Taste Deficiency' for Creatine"

The information presented above provides evidence for the first instances in which basic ideas, which later lead to the development of the concept of Biochemical Individuality, can be detected. Williams' (1931, p. 598) first public discussion of individuality as a scientific issue appears in an article spanning less than half a page in *Science*, in 1931. Using the term "Individual Metabolic Idiosyncrasies" for the first time, differences in human sensory factors are deliberated.[71] The article, entitled " 'Taste Deficiency' for Creatine" (Williams, 1931), is a summary of circumstantial observations made in connection with research on the substance creatine, first published in 1926 (Williams and Lasselle, 1926). Here, Williams describes his discovery that creatine, a substance usually described as bitter in

---

70 An example of such grouping can be found in *An Introduction to Biochemistry*, in which Williams explicitly includes man into the research of the nutritional requirements of mammals (Williams and Beerstecher Jr., 1948a, p. 229).
71 The term "Metabolic Idiosyncrasies" (also referred to as "MI" in this thesis) can be considered as synonymous to "Biochemical Individuality". MI is used on multiple occasions before the term BI is first published, most prominently in " 'Taste Deficiency' for Creatine" and *The Human Frontier* (Williams, 1931, p. 598, 1946a, p. 74).

textbooks, is entirely tasteless to himself and many other individuals (Williams and Lasselle, 1926). This work on differences in taste constitutes the first public mention of and explicit "research" on individuality in a scientific sense by Williams, though this small number of participants and lack of statistical analysis does not truly merit a description as a scientific trial.[72]

The above observations are not commented on or explained in the original 1926 publication; its statistical focus does not provide any further analysis. Five years later, Williams offers the omitted clarification, the apparent discrepancy of his findings to the data of traditional scientific literature inducing him to contemplate the issue further. Concluding that differences in taste perception could have an effect on an individual's affinity towards certain types of meat, Williams indicates how this variation may arise due to "individual metabolic idiosyncrasies" (Williams, 1931, p. 598). Lamenting the ignorance of the medical field toward human individuality, two further instances in which it may affect everyday life are elucidated. The medical phenomenon of adverse drug reactions, namely morphine and novocaine,[73] are described, alongside an additional instance of olfactory variation. Williams summarises the experience of an acquaintance pertaining to skunk odour, noting a peculiar lack of pungency-perception in this individual, deviating markedly from the stinging smell it has for most individuals (Williams, 1931, p. 598).[74]

More conclusive evidence for the relevance of "'Taste Deficiency' for Creatine" is provided by its reference in multiple of Williams' later publications (Williams, 1946a, p. 70, 1953a, p. 27, 1956a, p. X, 127). Alongside other "Miscellaneous Evidences of Individuality", this first small glance at the principle mechanisms and possible ramifications of BI is embedded alongside scientific studies dealing

---

72   This conclusion concurs with the findings of Davis et al., though their work quotes a different year (Davis et al., 2008, p. 10). Davis et al. (2008, p. 10) cite an article from 1928, "concerned with individual differences in the taste of creatine". However, this date is neither reflected in the list of publications provided by the institute, nor by the original publication (Davis, 2003b; Williams and Lasselle, 1926). The first article "The Identification of Creatine", upon which the 1931 analysis is based, was published in the *Journal of the American Chemical Society* in 1926 (Williams and Lasselle, 1926). When questioned about the subject via email on the 17.04.2020, Donald R. Davis could confirm he "feels confident that the 1928 date is a mistake", expounding further, that "it presumably refers to the 1926 paper with initial observations of taste differences to creatine".

73   See Section 4.3. Adverse Drug Reaction.

74   This example also reappears on multiple occasions (Williams, 1942a, p. 343, 1946a, p. 75, 1953a, p. 32, 1956a, p. x).

with individuality in taste (Williams, 1956a, p. 127). Williams (1956a, p. X) additionally describes how his "own particular interest in [Biochemical Individuality] probably stems from the laboratory observation, (…) that, although creatine was described by Beilstein as a bitter biting substance, it was found to be absolutely tasteless to many", in the preface of *Biochemical Individuality*. He additionally describes how the inability of some individuals to detect skunk odour played a role in impressing the relevance of variation (Williams, 1956a, p. X). These observations and comments in *Biochemical Individuality* clearly define "'Taste Deficiency' for Creatine" as the first publication meaningfully discussing the existence of Biochemical Individuality.

The use of the term "individual metabolic idiosyncrasy" furthermore constitutes a precursor to the term "Biochemical Individuality" in Williams' work. The idea of idiosyncrasy previously plays a role in Williams' notion of alternative reactions to medication,[75] while Metabolic Idiosyncrasy is later used synonymously to Biochemical Individuality (Williams, 1946a, p. 13). Both describe the concept of a metabolism specific to an individual human being. The term also appears in Archibald Garrod's (1857–1936) work on individuality, whose concepts are central to the development of BI (Garrod, 1923, p. 3).[76]

"'Taste Deficiency' for Creatine" serves the purpose of providing background information rather than presenting hard scientific facts. However, its relevance to the development of a concept of BI lies precisely in its generalist nature: it presents the anecdote, the by-the-by mention of a discovery which sparked interest and inspired further study. In his article, Williams (1931, p. 598) remarks how the field of metabolic idiosyncrasies "calls for extensive study", an undertaking he himself turns to in later years. "'Taste Deficiency' for Creatine" and its findings therefore represent a further step in the development of BI. It confirms the hypothesis of a pattern of coincidental findings leading to further understanding and depth of interest. It also proves Williams' increasing professional curiosity with regard to BI, as conclusively confirmed in an interview at the end of his career (Hodge and Williams, 1980a).

---

75 See Section 4.3. Adverse Drug Reaction.
76 The relevance of Archibald Garrod's work to the development of Biochemical Individuality is discussed in Section 6.4. "Distinctive Metabolic Traits".

4.6 Biochemical Individuality Before 1940

Metabolic Idiosyncrasies may be at the root of the differences in gustatory, sensory, and olfactory perception, as well as divergent reactions to certain medications.

4.7 Conclusion

The exact origins of Roger J. Williams' concept of Biochemical Individuality are difficult to place, due to the indiscriminate nature of his first accumulation of knowledge on the subject. Williams receives his first glimpses at interindividual variation while beginning, as all humans do, to explore the world around him throughout his childhood. These incidental findings and anecdotal evidence provide him with a first understanding of his own individualities. Sparking scientific curiosity, such first inklings are bolstered by technical evidence, gathered coincidentally during biochemical studies with yeast. The work on his first ever publication proves influential in hindsight, providing his first scientific observation of the existence of differences between strains of single-celled organisms. What can be determined with absolute certainty, however, is the first public manifestation of Williams' budding thought-process. "'Taste Deficiency' for Creatine", published in 1931, discusses "individual metabolic idiosyncrasies", clearly demarcating the first published description of biochemical differences between humans in his work (Williams, 1931, p. 598).

Prior to the publication of the first research purposely focused on the differences between human beings, the definition of BI is difficult to discern. The only piece of academic work on the subject is provided by Williams' publication in 1931. The abstract offered above is deliberately vague and suggestive, reflective of the formlessness that characterises the evidence available at this stage of BI's development. "'Taste Deficiency' for Creatine" makes attempts to explain interindividual differences; this thought process is not further reflected on in the works following its publication. While differences with relation to the growth of yeasts and other fungi appear in consecutive publications on Williams' research of vitamins, no similar discussions on or explanations for individuality can be found.[77] Williams' original thoughts on BI can be traced through the four instances portrayed in this chapter, indicating where his primary understanding of Biochemical Individuality originated.

---

77 See Chapter 5. Vitamin Studies.

5 Vitamin Studies

Though first allusions to Biochemical Individuality are made in and before the early 1930s, Williams' publications of the following decade do not elaborate on the concept. Though continually confronted with discrepancies in the nutritional requirements of yeast strains, only one study containing an explanation for these differences appears in the publications following "'Taste Deficiency' for Creatine" (Williams, 1931).[78] In 1933, Williams succeeds in determining the structure of pantothenic acid, an achievement garnering him fame and regard within the scientific community (Davis et al., 2008; Williams et al., 1933). Continuing his biochemical research on vitamins, Williams publishes his last article as a member of Oregon State College in May of 1939, before moving to Austin later in the same year (Williams, 1939a).[79]

Consequently, Williams' first published work from The University of Texas at Austin with Esmond Snell (1914–2003) appears in December of 1939 (Snell and Williams, 1939).[80] Having taken up a post as professor of biochemistry in Austin, Williams begins to revisit the field of individuality. Founding The Department of Biochemistry at The University of Texas at Austin in 1940, Williams' work markedly diversifies (Davis et al., 2008, p. 3). As Head of Department, he finds himself in a position to freely choose the topics of his research according to his own interests (Williams et al., 1966). Under his leadership, DBUT "consistently promot[es] independent thinking on the part of [its] members", encouraging all of its researchers to pursue any field in which they see potential (Williams et al., 1966, p. 3). This attention to and respect for the individual interests of his co-workers continues throughout his tenure at the university.[81] Accordingly, Williams' work of the following years is characterised by increased collaboration with other scientists and a diversification in research topics. The following chapter summarises the allusions to BI which appear in Williams' publications during

---

78 For further information, see Williams and Saunders (1934).
79 This definition has been made based on the list of articles provided by Davis (2003b).
80 Esmond Snell worked alongside Roger Williams at DBUT as a nutritional biochemist. He is credited with the discovery of folic acid and was later chair to the Department of Biochemistry at UC Berkeley (Sanders, 2003). For further information, see Sanders (2003).
81 See Section 16.1. Interview with Donald R. Davis.

the first half of the 1940s. Special focus is placed on those studies which reveal evidence of individuality in their research subjects.

5.1 Individuality of Yeast Strains

Though Williams' study of vitamins is not devised to detect individuality in the strains of yeast it utilises, multiple publications following "'Taste Deficiency' for Creatine" contain data indicating such variation. For example, Williams and Honn (1932, pp. 629, 632) discuss how "different strains of Saccharomyces cerevisiae require traces of different unknown substances for their growth stimulations" and that "some of the molds grew to greater volume before producing spores than did others". Additionally, the two scientists describe how "the molds as a whole were found to behave rather erratically in their growth" (Williams and Honn, 1932, p. 633), and that "apparently for carrying on the living activities of an individual fungus, certain chemical materials are indispensable" (Williams and Honn, 1932, p. 639). Though these allusions all indicate an appreciation of the individuality of different fungi regarding their growth behaviour and nutritional needs, no further elaborations are provided. Similarly, Williams et al. (1933, p. 2912) make reference to their attempt "to find 'other' strains of Saccharomyces Cerevisiae which might have simpler requirements", further insinuating individuality in their nutritional requirements. A third paper additionally "remark[s] that this particular strain of yeast probably grows more rapidly on a synthetic medium than any other yeast we have investigated" (Williams et al., 1933, p. 2913). None of these works contain statements regarding the possible relevance of such individualities, focusing instead on biochemical aspects of pantothenic acid.

Continuing research towards the nature of B vitamins, Williams and Saunders (1934) later publish the results of a comparative study of yeast growth, which contains further data of variation on a cellular level.[82] The evidence presented in Tables II, III, and IV of this publication clearly indicates an individuality in growth behaviour stimulated by multiple vitamins (Williams and Saunders, 1934, pp. 1890–1891). Though these effects are acknowledged, the analysis of the results is once more focused on the nature and relevance of the vitamins in question, rather than suggesting that these differences may be rooted in an

---

82 Donald Herbert Saunders was a colleague of Roger Williams' at Oregon State College. His name appears in two publications with Williams (Mosher et al., 1936; Williams and Saunders, 1934). No further information on his person is available at the time of writing.

individuality of the examined yeast strains. This article is, therefore, of relevance to the development of BI due to the fundamental research approach it presents, rather than the analysis it contains. Williams' previous method of research is largely based on studying identical factors in uniform research subjects. This kind of study, therefore, attempts to minimise individuality by design. A large study which directly compares the same factor for "different species of organisms", however, supplies information on the variation between these species and highlights such differences (Williams and Saunders, 1934, p. 1891). Though it does not provide the "repeated samples from the same well individuals collected under basal conditions" (Williams, 1956a, p. 4), which Williams later claims are essential for the research of individuality, this comparison of yeast strains differs distinctly from his previous publications as the study is designed to highlight these differences and provides evidence towards the aforementioned behavioural and nutritional individualities. Later studies of individuality also utilise the new approach taken in "The Effects of Inositol, Crystalline Vitamin B1 and 'Pantothenic Acid' on the Growth of Different Strains of Yeast".[83] This article, therefore, constitutes progress in Williams' research of individuality as his first study of this kind.

Mitchell and Williams (1940, p. 1535) similarly remark on the apparent "considerable variability in the amino-acid requirements and the synthetic abilities of various strains of yeast" in a study six years later. Though "The Importance of Amino-Acids as Yeast Nutrients" indicates an appreciation of interindividual differences, it neither delves into further research on this fact, nor attempts any explanation there for. Finally, Williams et al. (1940a, p. 1205) speak of "three distinct (and highly different) strains of '*Saccharomyces cerevisiae*' " in their study of yeast growth in various mediums. "The Relationship of Inositol, Thiamin, Biotin, Pantothenic Acid and Vitamin B6 to the Growth of Yeasts" further highlights the variability of yeast strains and discusses the differences in growth rate between the strains for different mediums (Williams et al., 1940a, pp. 1205–1207).

This lack of discussion regarding the individualities discussed above not only indicates Williams' differing research focus at the time, but additionally shows an absence of awareness regarding their importance. Scientific research often generates a range of values due to inexact measurements or a genuine variability of the objects studied. In his later research, Williams attributes great significance to the minute biochemical variations in humans and therefore actively seeks

---

83   See Chapters 7. The Scientific Study of Individuals and Alcoholism, 8. Genetotrophic Disease, 9. Human Individuality, and 11. Practical Genetotrophism.

out such evidence, be it in his own studies or the work of others. Here, without such explicit appreciation and interest, the individualities discovered are not attributed any further relevance and are therefore not deliberated on at length.

It cannot, however, be said that the variations in the growth behaviour of yeast strains is entirely lost on Williams and his colleagues. In a review of the vitamin research predating 1940, Williams (1941b, p. 53) explicitly dedicates one section to "Differences in 'strains' of yeast". Discussing the variability of "what has been considered one species", Williams discusses the "diametrically opposite" results gained from experimentation with various yeast strains (Williams, 1941b, pp. 53–54). These differences are used to indicate how laboratories' results could show wide variance in comparison to their colleagues' research, as well as how the debate over the indispensability of vitamins to yeast growth could develop. In accordance with this analysis, three figures showing markedly different growth patterns of three yeast strains are presented (Williams, 1941b, pp. 70–72). No allusion to any further relevance of these differences is made, clearly indicating a knowledge of the individualities of yeast strains and further underlining the hypothesis that these are simply not attributed any deeper meaning at the time.

Such glimpses of uncommented individuality, as well as Williams' clear knowledge of these differences, help to illustrate the development of BI and the changes in Williams' approach to research over time. This theoretical biochemical research includes no appreciation of the biochemical individualities of organisms, in spite of Williams' continual exposure to the facts of Biochemical Individuality.

5.2 The Vitamin Content of Tissues

While Williams' relocation from Corvallis, Oregon, to Austin, Texas, could be considered as the beginning of a new chapter, this change of setting does not translate to an immediate change of his research focus. In fact, there is no recognisable "break" when merely examining the list of his publications (Davis, 2003b). Unhappy with the leadership of the Department of Chemistry at OSC, Williams writes to his colleague and confidant Linus Pauling (1901–1994) about the ungratifying working conditions as early as December 1937.[84] Indicating his wish for a change of scenery, Williams discusses how his motivation to "sever

---

84 Williams and Pauling were in regular contact from 1936 onwards, discussing topics of research as well as personal matters. The entirety of their correspondence is held by the Special Collections and Archives Research Center of the Oregon State University Libraries and Press in Corvallis, Oregon. The correspondence files have kindly been made available in the form of digital scans for the purpose of writing this thesis. For

[his] connection with this institution" lies within the working environment as opposed to the promise of his research (Williams, 1937). His sympathies for his colleagues, and a wish to "clean things up" before leaving are indicated in a letter to Pauling after accepting his post at The University of Texas (Williams, 1939b). The continued collaboration between Williams and the department at OSC on the research of pantothenic acid further underlines this thought.

Continuing his work on the basic biochemistry of vitamins, some previous studies are upheld and the results thereof are released under the same naming system explicitly referring to Williams' collaboration with the Institute at OSC (Williams et al., 1940b, p. 1784).[85] While Williams' publications do not show an immediate change of direction, some of the ongoing research provides further evidence of the inherent individuality of organisms. Though none of this research is primarily conceptualised to study such individualities, the results of these works indicate the variation later central to Williams' concept of BI. Having previously concentrated on the chemical structure of pantothenic acid and other B vitamins, these studies deal with their significance and distribution. This constitutes a shift from theoretical basic research to the practical applications thereof. The illustration of the examples of individuality within this practical research, without an in-depth discussion thereof, further underlines Williams' lack of an attribution of deeper meaning regarding differences of yeast strains in his biochemical studies.

The first publication containing evidence of such aforementioned individualities is submitted for publication by Williams in February of 1940 (Snell et al., 1940b). Studying "The Effect of Diet on the Pantothenic Acid Content of Chick Tissues", Williams and his co-workers present comparative data for two groups of chicks. With one group receiving a diet supplemented with pantothenic acid and the other a control diet, the amount of pantothenic acid present in various tissues is documented and compared (Snell et al., 1940b, p. 560). The pronounced individualities of the chicks involved in the study become noticeable when regarding Tables II and III of the article (Snell et al., 1940b, p. 562). Though the study was conducted using chicks of identical age, the variation in weight and pantothenic acid content is not only marked between the two cohorts, but clear disparities

---

further information on Pauling's role in Williams' career, see Section 3.2 Academic Networks.

85 This is in reference to a succession of publications, all released under the title "Pantothenic Acid", followed by the corresponding numerals. All are included in the list of publications on the Website dedicated to Williams' publications (Davis, 2003b).

exist within both groups (Snell et al., 1940b, p. 559).[86] The discussion and summary of this research, however, remains very brief and provides little analysis for these values. The overall trend of the results yields that in "every tissue tested from chicks raised on the deficient diet, the pantothenic acid content was markedly lower than in the corresponding tissue from chicks fed the supplemented diet" (Snell et al., 1940b, pp. 563–564). Additionally, the pantothenic acid deviations from the control group are presented in percent. No comment is made on the interindividual differences of the chicks, and this article is not the only publication containing data which shows a high degree of individual variation without commenting upon this fact. Similarly, an article published in *The Journal of the American Medical Association* indicates a wide range of values in relation to human nutrition (Spies et al., 1940).[87] "Pantothenic Acid in Human Nutrition" compares the pantothenic acid content in blood samples of numerous individuals, as well as the subsequent changes in blood-levels of riboflavin following an injection of pantothenic acid. Significant variations in the range of ten percent and higher are indicated, though no relevance is attributed thereto (Spies et al., 1940, pp. 523–524). Later that same year, Williams is able to publish the chemical structure and successful synthesis of pantothenic acid with Randolph Major (1901–1976) (Williams and Major, 1940).[88] This doubtlessly constitutes one of Williams' most famous discoveries, cementing his name in the scientific world and public eye (The New York Times, 1940). He later receives the Chandler Award for Research Efforts for this discovery and is considered an expert in vitamins and nutrition following his work thereupon (The New York Times, 1942).

A final notable example of such studies comes in the form of the DBUT's first more comprehensive publication. Announced in *Science* in December of

---

[86] In Table II, the weight of the chicks receiving supplementation ranges from 147 to 174 grams and the weight of non-supplemented chicks ranges from 55 to 102 grams. Table III presents data in which the weight of chicks receiving supplementation ranges from 207 to 223 grams and the weight of non-supplemented chicks ranges from 73 to 101 grams. These values clearly indicate marked variability in the weight gain of chicks under identical circumstances (Snell et al., 1940b, p. 562).

[87] "*The Journal of the American Medical Association*" (also referred to as "*JAMA*" within this thesis) is a well-known international medical journal. The JAMA-Network encompasses 12 peer-reviewed journals and publishes weekly. For further information, see Bibbins-Domingo and Curfman (2023).

[88] Randolph T. Major was a Princeton-educated chemist and Director of Pure Research at Merck and Company in 1940. For further information on his life and work, see UConn Foundation (2023).

1940 (Williams, 1940), *Studies on the Vitamin Content of Tissues* volumes I and II are the first larger collections of research papers published by Williams and his colleagues at DBUT, and are released in 1941 and 1942 respectively (Williams et al., 1941, 1942).[89] Though only the second volume is referenced in *Biochemical Individuality: The Basis for the Genetotrophic Concept*, both provide evidence for the changing priorities of Williams' research.

Having discovered the structure of B vitamins, Williams and his colleagues develop techniques for their adequate detection in the form of assay methods. Such procedures are provided for riboflavin (vitamin B_2), pantothenic acid (vitamin B_5), biotin (vitamin B_7/H), nicotinic acid (vitamin B_3), pyridoxine (vitamin B_6), inositol, thiamine (vitamin B_1), and folic acid (vitamin B_9) in *Studies on the Vitamin Content of Tissues I* (Williams et al., 1941).[90] While "individual variations in the vitamin contents of different tissues" are mentioned, this first collection of studies explicitly does not aim to present "any extended interpretation of the findings here set forth" (Williams, 1941c, p. 9). Referring to the lack of sufficient evidence to extrapolate from and interpret the data adequately, no such discussions appear.

Throughout the entire publication, merely two further references with regard to the aforementioned disparities are made. Both appear in the article "The 'B Vitamins' in Normal Tissues (Autolysates)" and remain ambiguous (Wright et al., 1941, pp. 57, 58). The first relation pertains to "similarities and differences" between species, merely mentioning the findings that vitamin concentration in tissues seems to be negatively correlated with the size of the organism in question (Wright et al., 1941, p. 57). Furthermore, the possibility of differences between the sexes is posed, though the available assay methods do not provide the necessary efficiency and exactitude needed for statistical analysis (Wright et al., 1941, p. 58). In light of Williams' later research focus, such evidence is highly relevant.

As above, the significance of this source lies within its tables and figures, as "The 'B Vitamins' in Normal Tissues (Autolysates)" includes no further discussion of the values presented. Readers must, therefore, make their own

---

89 Williams announces application of the assay methods developed in his institute to "animal tissues of numerous types, including embryonic tissues and tumors" (Williams, 1940, p. 579).

90 In 1941, inositol was assumed to constitute a vitamin for human beings, as it acted as such in rodents and yeast. With advancing research into the biochemical pathways of human physiology, it is now known that the human body is capable of producing inositol from glucose (SRI International, 2014, 2023). It is, therefore, no longer classified as a vitamin for humans.

assumptions and draw their own conclusions from the offered data. The values depicted in Table 9 and Figure V represent statistical averages, with Table 9 additionally presenting the maximum and minimum values documented for concentrations of B vitamins in various tissues (Wright et al., 1941, pp. 40, 47). Merely providing mean values, as well as the maximum and minimum values for quantities collected from multiple samples, the possibility of meaningful statistical analysis on the part of the reader is highly limited. The presented data does, however, allow for the calculation of range values, which can provide an indication for the spread of values.[91] When these are calculated, multiple instances with inordinately large values stand out.[92] Importantly, the number of samples is not large enough to produce values from which one may reliably extrapolate for an entire population.[93] This could be a viable explanation for the absence of further analysis. While statistically inconclusive, these values do provide evidence of the first cross-sectional study of individual complex organisms carried out in Roger Williams' laboratories. The lack of discussion, possibly due to a small sample size, indicates the difference in focus described previously.[94]

Though "The Effects of Inositol, Crystalline Vitamin B1 and 'Pantothenic Acid' on the Growth of Different Strains of Yeast" provides insight into differences in

---

91 The range of a set of values is defined as "the largest minus the smallest of a set of variate values", and therefore "may afford a reasonable estimate of the population standard deviation" (OECD, 2008, p. 448). As "The 'B Vitamins' in Normal Tissues (Autolysates)" merely supplies mean values, it is impossible to calculate the standard deviation or variance (the statistical values usually calculated to portray variation) for this data set.

92 Three examples, all taken from "The 'B Vitamins' in Normal Tissues (Autolysates)", are presented below:
1.) Nicotinic acid content of rat spleen:
Mean = 240, Maximum = 390, Minimum = 130, Range = 260
2.) Pantothenic acid content of rat liver:
Mean = 370, Maximum = 670, Minimum =230, Range = 440
3.) Pantothenic acid content of rat spleen:
Mean = 59, Maximum = 110, Minimum = 30 Range = 80
All values are given in γ/gram (Wright et al., 1941, p. 40).

93 The results presented by Williams and his colleagues are based on samples of between one to six test subjects, as can be derived from Table V and Figure 9 of "The 'B Vitamins' in Normal Tissues (Autolysates)" (Wright et al., 1941). A discussion of the calculations for the minimum size of such a study would exceed the scope of this thesis. For further information, see Machin et al. (1987) and Röhrig et al. (2010).

94 Section 5.1. Individuality of Yeast Strains.

growth behaviour of yeasts at an earlier juncture, this research on individual mammals signifies a further step. As BI is later applied to humans, the progress from comparative studies on single-celled organisms to more complex mammals is significant. Additionally, Williams' own later research into BI is largely based on experiments with rats.[95]

Finally, Williams' announcement of these publications in *Science* offers additional evidence for the relevance of these works. A study of "the relationship of these results to the problems of metabolism" is explicitly publicised, though it adds the condition that only "thorough study can yield results which are significant and capable of interpretation" (Williams, 1940, p. 579). As has been previously discussed, the size of cohorts in these studies is not large enough to provide statistical significance, which explains the omission of discussion towards the applications of their results to "problems of metabolism" (Williams, 1940, p. 579).

Williams' collaboration on the vitamin content of animal tissues with his colleagues in Austin represents the first evidence of a cross-sectional study of complex organisms at DBUT. The discussed works provide evidence for the type of individualities unveiled by Williams' vitamin research in Texas. This continual exposure to evidence of BI finds significance in the practical applications of vitamin research. Therefore, these studies represent important evidence for the evolution of Williams' and his colleagues' research on vitamins, from a solely biochemical perspective to a broader concept of their practical relevance to organisms in general. This concept is later reflected in the nutritional aspects of BI.[96]

5.3 Practical Applications of Vitamin Research

The sources previously examined are essentially scientific in their focus. The information they provide, while relevant to the scientific community and vitamin research in general, is of little practical use to the wider public. As the articles critically offer no analysis of the possible practical applications of the information they provide, there is little possibility for the utilisation of the data they present outside of further research. Two later articles, one published in *JAMA* (Williams, 1942b) and the other in *Science* (Williams, 1942a), indicate such possible future usefulness. Published in April and May of 1942, these articles not

---

95   See Chapters 7. The Scientific Study of Individuals and Alcoholism, 8. Genetotrophic Disease, 9. Human Individuality, and 11. Practical Genetotrophism.
96   See Sections 8.1. The Metabolic Individualities of Rats and 9.3. *Nutrition and Alcoholism*.

only summarise the research on vitamins predating their respective publication, but also provide an outlook on potential applications of their data and new fields of research as a result therefrom.

In the opening remarks of the first of these articles, "The Approximate Vitamin Requirements of Human Beings", Williams brings to light the true directive behind his vitamin research before 1942: "Most of the research on vitamins and nutrition is directed ultimately to the solution of problems of human nutrition" (Williams, 1942b, p. 1). Predating the publication of this article, the motivation behind Williams' chosen field of research remained unknown. As Roger Williams' brother Robert R. Williams' (1886-1965) is credited with the discovery of thiamine, and with this the origin of the nutritional disease Beriberi, this potential application of Roger Williams' vitamin research is close at hand (Williams, 2013).[97] Devised to "point out that the approximate human requirements for various of the B Vitamins can now be estimated with some reliability", Williams (1942b, pp. 6-7) presents an analysis of the cumulative results of his prior vitamin research.

Additionally, the historical context of "The Approximate Vitamin Requirements of Human Beings" – and the analysis of the approximate required quantities of vitamins by human beings contained therein – is vital to its adequate appreciation: "[i]n a war situation such as ours it is particularly desirable that we be able to apply whatever we may know or learn to practical ends" (Williams, 1942b, p. 1). The United States having joined the Second World War in December of 1941, Williams portrays how the question of the ideal nutritional equipment of American soldiers becomes a new and pressing aspect of his research.[98] All previously discussed sources (with the exception of *Studies on the Vitamin Content of Tissues II*) predate the direct involvement of American troops in World War II. Therefore, this source is of additional value, as it shows the effects of exogenous pressures on scientific research in the United States, including that of Roger Williams.

To this end, the article in question provides approximated values for the daily

---

97 Robert R. Williams Jr. is best known for his isolation of crystalline thiamine and his work discovering the cure for Beriberi. Beriberi is a disease of the nervous, muscular, and cardiovascular systems arising from malnutrition. Vitamin B_1 (Thiamine) plays a central role in the aetiology of this disease. For more information on the discovery of thiamine and its applications to Beriberi, see Williams (2013). Further information on Robert R. Williams role in his brother's career is previously discussed in Section 3.2 Academic Networks.

98 For further information of the History of WWII, see Keegan (2005).

intake of thiamine, nicotinic acid, riboflavin, pantothenic acid, biotin, inositol, pyridoxine, and folic acid necessary for the healthy nutrition of a human adult. Additionally, values for pregnant and lactating women are supplied (Williams, 1942b, pp. 4–5). The concepts of a "well rounded" diet (Williams, 1942b, p. 1), and that "most people do not get as much of the various B vitamins as they probably should have" (Williams, 1942b, p. 5), play a vital role in Williams' later "Genetotrophic Concept".[99][100] Crucially, Williams does not scrutinise the medical ramifications of this vitamin deficiency, which are central to GC, in "The Approximate Vitamin Requirements of Human Beings" (Williams, 1942b). It is, however, Williams' first mention of such vitamin deficiencies in a non-supplemented, regular diet. Furthermore, Williams speaks of the weaknesses of extracts or preparations, which are deemed "wholly inadequate as sources of the vitamin B complex" (Williams, 1942b, p. 6), and the "present economic waste involved in the production and sale of preparations of questionable value" (Williams, 1942b, p. 7). This critique of the inadequacies of the supplementation industry is additionally noteworthy, because much of the work of Williams' later career is based on the role of nutrition in the prevention and curing of various diseases.[101] Here, the correct supplementation of vitamins and nutrients is of absolute essence. Holder of a patent for the *Production of Pantothenic Acid and Other Related Growth Promoting Substances* (Williams, 1947a), Williams was accused of having ulterior, monetary motives when recommending certain supplements later in his career (Cotlier, 1977). These allegations were rebuked by his colleague Donald R. Davis (born 1941) and there is no evidence toward their truthfulness (Davis, 1978).[102][103] This prior statement from 1942 additionally indicates contempt for the supplementation industry.

---

99 "Genetotrophic Concept" and "Genetotrophic Principle" are synonymous and are also referred to as "GC" within this thesis.
100 See Section 8.1. The Metabolic Individualities of Rats.
101 See Chapters 7. The Scientific Study of Individuals and Alcoholism, 8. Genetotrophic Disease, 9. Human Individuality, and 11. Practical Genetotrophism.
102 Donald R. Davis is a retired chemist who worked with Williams at The Clayton Foundation Biochemical Institute as of 1974. He originally was a visiting researcher at the institute on a sabbatical leave in 1973, returning the following year to join the institute. Having worked closely with Williams up until his death in 1988, Davis stayed on at the institute until 2007 and still manages a website dedicated to Roger Williams' research. For further information, see Section 16.1 Interview with Donald R. Davis.
103 In personal correspondence on the 24.04.2019, Mr Davis describes Williams' relationship with supplements and their producers further: "Williams carefully tried to

Though a further article, "Vitamins in the Future", also provides an abridged summary of predating vitamin research, its focus, as suggested by its title, lies on the future applications of the research described (Williams, 1942a). Possible applications of knowledge toward chemotherapy, cancer research, psychological and behavioural differences on the basis of diet, and the fostering of "higher intelligence" and morality are discussed (Williams, 1942a, pp. 343–344). A substantial section of this piece is additionally devoted to the question of "individual differences" amongst humans (Williams, 1942a, p. 343):

> How great a contrast is there, biologically speaking, between an inbred colony of experimental animals and, say, the population of New York City, where even within each of the numerous racial groups there are tremendous genetic differences. But our nutritional knowledge when applied must be used in precisely such diverse groups. It may be that some day the medical profession will be able to concentrate its attention upon the very thing that the nutritionist likes to eliminate as completely as possible, namely, the variation in the needs of individuals. (…) Not only the heritage of an individual but his case history may conceivably make for altered and probably increased vitamin requirements. (Williams, 1942a, p. 343)

In this excerpt, Williams discusses the idea of individualities in human genetic makeup more than a decade before the structure of DNA is established.[104] Additionally, he claims that such differences occur not only between ethnic groups, but that differences also occur on an individual level within these. This can be seen as an extension and further evolution of the idea of differences between strains of yeast or rats previously discussed.[105] Furthermore, a medical appreciation and consequent therapy on the basis of these genetic differences is suggested. Such decisive statements are not characteristic of Williams' work until this point. In the following, Williams provides further anecdotal evidence, some of which having appeared in previously cited sources:

---

avoid any appearance of profiting from nutritional supplements. For that reason, he delayed publication of one of his books (probably Nutrition in a Nutshell, 1962) until after his patent on the synthesis of pantothenic acid had expired. As I recall he received a small portion of the royalties before expiration. (…) In the late 1970s when he ordered a personal supply of the 'nutritional insurance' supplement and the company did not charge him, he asked me to buy subsequent supplies for him".

104 The structure of DNA as a double-helix was proposed by James Watson (born 1928) and Francis Crick (1916–2004) in 1953. For further information, see Watson (2010).

105 See Sections 4.2. "The Vitamine Requirement of Yeast" and 5.2. The Vitamin Content of Tissues.

> We know that the chemistry of our individual bodies is not all exactly the same, otherwise a bloodhound could not use his nose to distinguish between individuals. It is a well-known fact, though not always recognized in practice, that individuals do not all respond alike to common drugs. I once had a student who had his tonsils removed almost without an anesthetic, because the operating physician could not believe that he was unaffected by novocaine, even though the fact had been demonstrated previous to this occasion. Curious individual peculiarities sometimes show themselves. I have an acquaintance who, though his sense of smell is normal in all other known respects, is unable to detect the odor of a skunk. For him, the pure substance n-butyl mercaptan, the active principle of "skunk perfume", has no striking or obnoxious odor. When such remarkable differences exist with respect to other chemical substances it would not be surprising if the vitamin requirements of some individuals deviated sharply from the mean. Virtually nothing is known at present regarding this possibility. (Williams, 1942a, p. 343)

The nature of the anecdotes, which remain uncited, has many similarities with later publications.[106] The first sentence of this section is, however, also of most poignant relevance. The fact that the chemistry of every individual human being is unique constitutes a central argument in the concept of BI. Such a public proclamation of this fact by Williams is of importance because it provides the first evidence of Williams accepting and promoting such an assumption.[107] Offering day-to-day examples to provide evidence for his claim, similar types of which are discussed previously,[108] Williams highlights the need of further research on the individualities in his chosen field of research, namely the individualities of vitamin requirements. These are further discussed in "The Approximate Vitamin Requirements of Human Beings" (Williams, 1942b). Having discussed the possibility of "individual metabolic idiosyncrasies" in 1931 (Williams, 1931, p. 598), Williams here utilises the same knowledge in his vitamin research and further applies this thought-process to the possibility of medical relevance.

Finally, Williams mentions individualities between the nutritional requirements of humans in the context of the difficulties they bring to the research of ideal nutrition in "The Approximate Vitamin Requirements of Human Beings":

> A fully adequate idea of the requirements of human beings for the various vitamins could presumably be obtained only as a result of a series of extended controlled studies using human subjects. Even if it were feasible to plan and carry out experiments of this type just as animal experiments are planned and carried out, individual differences,

---

106 See Section 6.4. "Distinctive Metabolic Traits".
107 See Section 6.4. "Distinctive Metabolic Traits".
108 See Chapter 4. Origins of Biochemical Individuality as a Concept.

assuming that small numbers of subjects were used, would doubtless make the results very irregular. (Williams, 1942b, p. 1)

While the work previously cited supplies evidence of individualities in microorganisms and more complex mammals, the excerpts discussed in this section provide Williams' first application of this knowledge to humankind. These concepts constitute the cumulative results of his prior research. These articles can be identified as the first publications with true relation to BI in its later form. The previously referenced sources, relating to the individualities of less complex life forms and mammals, provide points along a learning curve. "Vitamins in the Future" and "The Approximate Vitamin Requirements of Human Beings", however, encompass these with regard to human beings (Williams, 1942b, 1942a).

This does not mean, however, that Williams' focus deviates entirely from the biochemical research his career is based on up to this point. The years after 1942 are characterised by a mixture of "hard" biochemical research and the "softer" nature of popular science publications. In fact, the greatest part of Williams' publications remain distinctly biochemical in their focus up to the publication of *The Human Frontier* in 1946 (Williams, 1946a).[109] Articles on the distribution of vitamins in the tissues of various mammals, insects, and fungi and the biochemical structure and significance of folic acid are exemplary for his continued scientific focus (Mitchell and Williams, 1944; Williams, 1943a). Within these scientific articles, the variations of vitamin distribution between species and strains of the same species continue to become apparent (Williams, 1943a, p. 234). Having amassed a high regard within the scientific community through his research on vitamins, Williams' research increasingly begins to develop towards other fields in the following years. Indicated by his inclusion in multiple publications on the "who's who" of science, this fame highly benefits his later public education books (Cook, 1943; Dickson, 1944; Faller, Undated; McKeen Cattell, 1943; Nichols, 1943; Rocker, 1943).

It is worthy of mention that research outside of the practical applications of vitamins is simultaneously underway at DBUT. Significantly funded by the Clayton Foundation for Research as of 1940, the field of cancer research is also of interest to Williams and his co-workers. In fact, Williams claims at a later point that "the deepest roots of the Institute go back to the joint interest of Mr. Benjamin Clayton and myself in the fundamentals of the cancer problem" (Williams et al., 1966, p. 2). Though principally focused on pantothenic and folic acid, Williams' involvement in multiple publications on the subject of cancer can

---

109 See Chapter 6. *The Human Frontier*.

be noted (Pollack et al., 1942b, 1942a, 1942c; Taylor et al., 1942c, 1942a, 1942b; Taylor and Williams, 1945; Williams et al., 1945). These do not pertain to BI and do, therefore, not merit in-depth discussion within the scope of this dissertation. Such cancer research at DBUT progresses into the 1950s, though Williams' involvement therein cannot be found after 1945 (Taylor et al., 1953).

5.4 What to Do About Vitamins

Having amassed considerable knowledge on the biochemistry of vitamins, Williams becomes more outspoken on subjects outside of his specific realm of research. Furthermore, his publications grow more prominent in the following years. This increased expertise and growing reputation is demonstrated by his nomination to author multiple reviews on the general state of the research in his field (Williams, 1943b, 1943a). These go beyond the studies which he himself published and represent the entirety of the knowledge of vitamins in 1943. The publication of the second edition of his biochemical textbook in 1948 further underlines his status as an expert within his field, though this topic is discussed elsewhere (Williams and Beerstecher Jr., 1948a).[110] In addition to the aforementioned diversification, Williams begins to direct his focus onto the underlying political issues of his field. Lamenting the underrepresentation of chemists in *American Men of Science* of 1944 (Cattell, 1944), he comments on chemistry's deficient status within the scientific community (Williams, 1945b). Calling upon the large number of registered chemists within the "National Roster of Scientific and Professional Personnel" (American Association for the Advancement of Science, 1942), Williams laments the fact "that chemists have not been good salesmen or advertisers" and that therein lie the origins of the lack of regard for chemists and their consequent underappreciation. When regarding the list of Williams' publications prior to 1944 with this critique in mind, a similar lack of advertisement for his own research and work becomes apparent (Davis, 2003b). It is, therefore, not surprising that Williams' first publication for laymen appears in the year following this epiphany.

What to Do About Vitamins, published in 1945 and written as a "diversion" during a month of vacation (Williams, 1945c), is Williams' first publication which can be characterised as "health education" for the general public (Williams, 1945a).[111] All previously published works, including all but one of the textbooks

---

110 See Section 7.4. An Introduction to Biochemistry, Second Edition.
111 The World Health Organisation defines "Health Education" as "any combination of learning experiences designed to help individuals and communities improve their

discussed elsewhere,[112] require a certain level of chemical or more specialised knowledge to be fully understood.[113] The university textbook *Introduction to Biochemistry* presumes, "that the student using this text will have had a substantial grounding in the field of chemistry, including at least a year course in organic chemistry, and that he will not start the study of biochemistry before the senior year", and thus reflects the nature of Williams' other works (Williams and Beerstecher Jr., 1948a, p. iii). None of his articles contain explanations for the basic (bio-)chemical principles their research is founded on. As all articles are published in specialised journals and the presumed audience is of a professional scientific nature, this is to be expected.[114] *What to Do About Vitamins*' audience, however, is the "typical American" (Williams, 1945a, p. v). Tables, graphs, and charts are explicitly produced to be easily understood (Williams, 1945a, loc. Back flap). The 56-page book constitutes an abridged summary of the knowledge gained from Williams' vitamin studies, condensed for public appreciation. The language used in this publication is kept simple, with vitamins referred to as "lubricants" and compared to motor oil in an easily accessible metaphor (Williams, 1945a, Chap. 3). This change in language is common to all of Williams' work for laypeople, altering or explaining technical terms to furnish further understanding (Hodge and Williams, 1980a).

The choice of publisher for this book, however, calls the previous assertions into question. Unlike Williams' later books aimed at the general public, *What to Do About Vitamins* is published by a university publishing company, The University of Oklahoma Press (Williams, 1945a).[115] A university press company, due to its nature of academic publishing, is less available to the general public than non-university publishing houses. This is also indicated by the relatively small number of sold copies, counting merely 2735 in 1950 and 3007 in 1953 (University of Oklahoma Press, 1950, 1953a). This choice of publisher, therefore, diverges from the general tone of the book.

---

       health, by increasing their knowledge or influencing their attitudes" (Baumann and Karel, 2013).

112 *Introduction to Organic Chemistry* is aimed at less advanced students and therefore conveys the basic understanding needed for further scientific study (Williams, 1935a).

113 See Section 7.4. An Introduction to Biochemistry, Second Edition.

114 For a complete list of these publications, see Davis (2003b).

115 See Section 6.2. Audience.

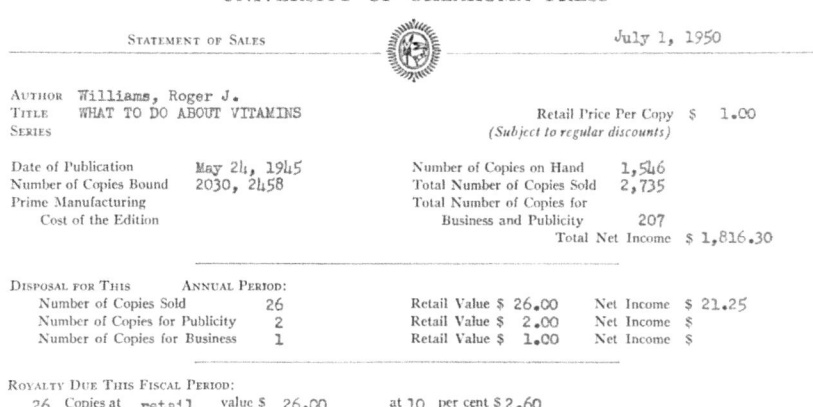

Fig. 3: Statement of Sales by University of Oklahoma Press on 01.07.1950 indicating that *What To Do About Vitamins* had sold a total of 2735 copies; Dolph Briscoe Center for American History, The University of Texas at Austin, Roger Williams Papers, Box 88-087/4

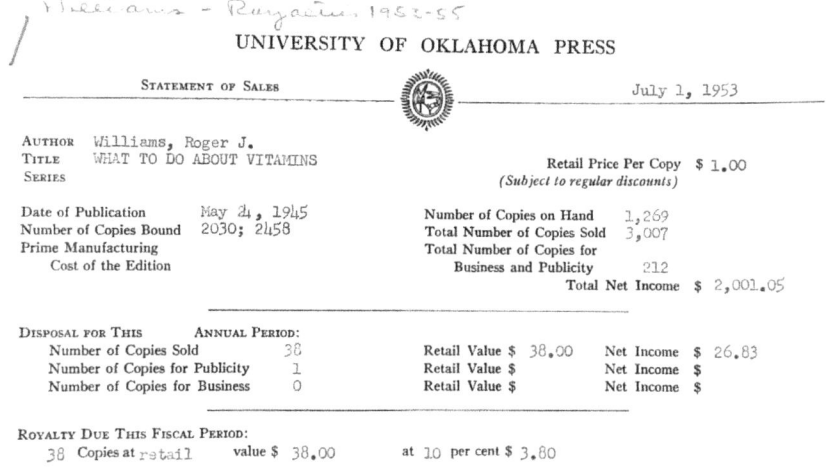

Fig. 4: Statement of Sales by University of Oklahoma Press on 01.07.1953 indicating that *What To Do About Vitamins* had sold a total of 3007 copies; Dolph Briscoe Center for American History, The University of Texas at Austin, Roger Williams Papers, Box 88-087/4

As stated above, the principle purpose of *What to Do About Vitamins* is to present "the means of obtaining vitamins, as well as the facts and principles underlying our need for them (…) so that the reader may develop expertness in the choice of foods and in deciding when and if vitamins should be purchased" (Williams, 1945a, p. V). This culmination of vitamin research, however, merely contains a singular reference to individualities in relation to vitamin requirements. With regard to the exact amounts of vitamins required for "best results in the human body", Williams states, that "no one knows precisely what amount of each of the lubricants is required (…) For one reason we doubtless have individual differences – we do not all require the same" (Williams, 1945a, p. 29). The overall context of this work and its mode of publication makes this statement – though its message is also reflected in previous publications – relevant to the development of BI. Williams' works on BI are largely aimed at creating an understanding in the general public rather than focussing purely on comprehension by the scientific community.[116] It can, therefore, be described as popular science.

The assertion of the idea that every human has specific individual nutritional requirements is later central to the Genetotrophic Concept. It bears relevance because it introduces an aspect of BI, however vaguely, to the general public and therefore constitutes the first example of public health education with an appreciation of BI by Williams. The importance of this excerpt should, however, not be overstated in that its importance to the readership or public at large will not have been great. It constitutes one sentence in a well-received book and does not feature in any reviews of the book found at the time of writing (Books Received, 1945; Good Things To Know About Vitamins, 1945; The Physician's Bookshelf, 1945; Vitamins and Their Roles in Everyday Diet, 1945; What to Do About Vitamins, 1945; What To Do About Vitamins, 1945a; What To Do About Vitamins, 1945b; What to do About Vitamins, 1945; What To Do About Vitamins, 1946; Davidson, 1945; Frederick, 1945; Glass, 1946; Stafford, 1945). Yet this first indication of BI signifies a further step in the development of the concept, continued by the publication of *The Human Frontier* in the following year (Williams, 1946a). It provides evidence for a change of mindset with regard to the importance of a public appreciation of (bio)chemical research, as well as increasing Williams' notoriety in the scientific and lay community. While previously focused on purely scientific appraisal, Williams increasingly addresses the general public through his publications from this point onward. The potential and solutions that chemistry and biochemical research have to offer society as a

---

116 See Sections 6.2. Audience and 7.1. "Biochemical Individuality and Its Implications".

whole become a central theme of Williams' publications. He later highlights the importance of public support and encouragement from the general population and the relevance of cultural change to the acceptance or dismissal of scientific theories in an interview at the end of his career (Hodge and Williams, 1980c). *What To Do About Vitamins* is Williams' first attempt at gaining public appreciation for his theories and research.

Though BI becomes an increasingly important topic in Williams' work, this interest in and appraisal of human nutritional variation is not immediately reflected by all of his publications on nutrition. Contributing to Michael Wohl's (1889–1970) book *Dietotherapy: Clinical Application of Modern Nutrition*, Williams speaks of the possible clinical applications of pantothenic acid (Williams, 1945d).[117] Here, Williams does not discuss human individuality as a root of "deficiency not due to dietary lack", instead focusing on the possibility of infection or metabolic disturbance leading to pantothenic acid deficiency unrelated to its intake. Again, this exemplifies that not all of Williams' work is characterised by a search for, and allusion to, individuality. It demonstrates the lack of clear proof and conviction regard the concept of BI which characterises the first half of the 1940s. *What To Do About Vitamins* is therefore not a consequent turning point in the development of BI, though it doubtlessly paves the path for the publication of Williams' next important step: *The Human Frontier*.

5.5 Biochemical Individuality Preceding *The Human Frontier*

The vitamin requirements of individual human beings differ markedly, much as every individual is unique in a wide variety of aspects.

5.6 Conclusion

The research of Roger Williams and his colleagues on the biochemical properties and distribution of various vitamins is, while not principally concerned therewith, of significance to the development of BI. In "Vitamins in the Future", Williams provides further justification for the extensive discussion of his principle research on vitamins in this thesis: "These questions could never have been *asked* were it not for the knowledge already gained through extensive research" (Williams, 1942a, p. 340). All extrapolations from and applications of

---

[117] Michael G. Wohl was an medical doctor and associate professor of medicine at Temple University School of Medicine (Dietotherapy, 1945). For further information, see Shuman (1971).

the research of vitamins, to which aspects of BI can undoubtedly be counted, would be impossible without this principal research. Initiated by the discovery of individualities in yeast strains and the study of the vitamin content of tissues, the importance of the first manifestations of BI in Roger Williams' research and published articles is great.

6 The Human Frontier

The Human Frontier: A New Pathway for Science Toward a Better Understanding of Ourselves (Williams, 1946a),[118] published in 1946, constitutes a noticeable divergence from Williams' earlier uniform mode of publication (Williams, 1946a). All articles published prior to *The Human Frontier*,[119] with the exception of "Achieving Full Employment After the War" (Williams, 1945e),[120] appear in scientific journals and are of a technical and biochemical nature (Williams, 1946a). *The Human Frontier*, however, discusses individuality with regard to its implications on "social science" (Williams, 1946a, p. 3), demonstrating the necessity for a "science of human beings" (Williams, 1946a, p. 5). Williams (1946a, p. 5) suggests this field of research be named "humanics", further exploring the ramifications of human individuality on society with regard to psychology, religion, education, marriage, criminology, medicine and medical research, leadership, environment, employment, and international relations (Williams, 1946a Table of Contents). Additionally, "charlatanism in politics and elsewhere, (…) alcoholism, group bigotry (whose name is legion) and war" are discussed (Williams, 1946a, p. 5). Williams later states that *The Human Frontier* is his first meaningful publication on individuality (Hodge and Williams, 1980d). Though too extensive to merit meaningful analysis of this publication within the constraints of this thesis, relevant allusions to and examples of Biochemical Individuality will be discussed, with special focus placed upon those also appearing in *Biochemical Individuality: The Basis for the Genetotrophic Concept*. The audience and raison d'être of *The Human Frontier* and the issues surrounding social control are additionally analysed, while reviews are considered in order to indicate the book's reception within the academic community.

---

118 "*The Human Frontier: A New Pathway for Science Toward a Better Understanding of Ourselves*" is also referred to as "*The Human Frontier*" within this thesis.
119 For a full list of publications, see Davis (2003b, 2003c).
120 Published in *Science*, Williams' article suggests it could be "*profitable* and *useful*" to employ and invest more in the field of research following the end of the Second World War (Williams, 1945e, p. 537).

6.1 "Humanity Must Understand Itself"[121]

Williams describes the rationale for *The Human Frontier* in its first Chapter "I. Why the Science of Humankind?" (Williams, 1946a, pp. 3–19). Here, an account of prolonged wonderment with regard to the central problems of social science is given, concluding that "to make human beings better known – to find out as completely as may be possible how and why they behave as they do" is essential hereto (Williams, 1946a, p. 3). According to Williams, the lack of understanding and appreciation for individuality in the early 1940s stems from an absence of scientific interest and thorough investigation. Additionally, he presents a viewpoint which would later become central to his work on BI.[122]

> Those who have dealt with the scientific study of man most intensively have had little or no concern for the possible social implications of their investigation. Man has been studied in pieces and not in his entirety, and we have been so devoted in our scientific work to the biological robot, *man-in-the-abstract*, that much of our knowledge is of very limited value from the social standpoint. Society can by no means be dealt with as though it were made up of individuals who are all alike, and yet this scientifically untenable conception is the basis of a large part of our social thinking and acting. (Williams, 1946a, p. 4)

The issue of the identical treatment of all human beings in lieu of a personalised approach is presented as fundamental to social science and the problems of society in general. The only solution for these, according to *The Human Frontier*, could be a science focussing on these differences that scientific research had, to this point, largely ignored.

> *A science of human beings is essential in an age of science.* (…) This science of human beings (which has for its purpose improvement in social control) we may call *humanics*.
> * Only by learning its basic truths, teaching them to our youth, and by extending greatly the boundaries of our knowledge can we cope with numerous social problems (…).
> *This rarely used word is defined as the "study of human nature" (*Webster's New International Dictionary*), and is parallel to mechanics, dynamics, acoustics, statistics, etc. (Williams, 1946a, p. 5)

---

121   This quote is taken from the last page of *The Human Frontier* (Williams, 1946a, p. 301). To improve the clarity of this section-title and avoid confusion, the citation for the quote is presented in this footnote, deviating from the citation style used in the rest of this dissertation. The capitalisation of this quote has been altered to fit the style of section-titles of this dissertation. The original quote reads "*Humanity must understand itself*".

122   Here, the focus merely lies on the meaning of these individualities for society, while BI additionally relates to the possible medical ramifications thereof.

The concept of "humanics" as a holistic approach to studying mankind becomes crucial for the concept and research of BI. This critique, that "*the full searchlight of science has never in any instance been turned on a real living individual, living or dead*" is critical to defining Williams' motivation for the publication of *The Human Frontier* (Williams, 1946a, p. 11). His proposal of studying human beings in a new and specialised science is therefore fundamental to *The Human Frontier*. This research, to be conducted by experts from multiple different fields in unison, is suggested to fuse the work of physiologists, biochemists, psychologists, sociologists, and many more (Williams, 1946a, p. 169).

The stylisation of this new science as a panacea for "combatting the evils of society" is additionally noteworthy (Williams, 1946a, p. 6). It indicates an absolute conviction and belief in the idea. The postulation that humanics could "set us free" by supplying "the *truth* about ourselves" further indicates unconditional confidence (Williams, 1946a, p. 6). The possibilities for betterment through humanics are suggested to be legion: success of political leaders, a better environment of physical, psychological, and social development, an educational revolution, reduction of lawbreaking, psychiatric disease and general frustration, and improved choice of job and martial partners are all presented as issues easily solved by following Williams' ideas (Williams, 1946a, pp. 6–8).[123] Built on "a scientific basis – for tolerance and good will", Williams' plans suggest the possibility of "rais[ing the individual's] position to one of dignity and honor" (Williams, 1946a, p. 8). Even the establishment of a "world government" is presented as a possibility (Williams, 1946a, p. 8). These glowing predictions of grandeur and utopian societies built on tolerance further indicate the importance attributed to the concept of humanics.[124]

Furthermore, Williams reveals his belief in a maladministration regarding contemporary knowledge of individuality at the time. Additional reasoning for the publication of *The Human Frontier* becomes evident at the end of the book's first chapter:

---

[123] Here, similarities with "Vitamins in the Future" become apparent, in which the topics of psychology, behaviour, intelligence, and morality are discussed. See Section 5.3 Practical Applications of Vitamin Research for further details.

[124] As Williams' purely biochemical background means he is inexperienced in social science and publication therein, this glowing description of his suggested solution to essentially all pressing societal problems could be attributed to an absence of extensive experience in the field. His descriptions are largely seen to oversimplify highly complex issues. For further discussion, see Section 6.6 Reviews of The Human Frontier.

Above all, we must see the urgency of the task; a new science must be developed – one which will concentrate on the comprehensive scientific study and understanding of actual human beings, such as those represented by ourselves, our neighbors, associates, friends, and enemies. (Williams, 1946a, p. 18)

The ambition for this publication is consequently not only the accurate description of a deficient condition, but the substantial reorganisation and reconsideration of pre-dating scientific approaches. Much of Williams' later works, including *Biochemical Individuality: The Basis for the Genetotrophic Concept* (Williams, 1956a), act as an appeal for the rethinking of human research, though the solutions he suggests to solve these issues change over time.[125] *The Human Frontier* follows a similar goal (Williams, 1946a). This becomes apparent by the proclamation that this science of humanics "should be of the nature of an *applied* science – one that is developed because of the practical service it will render (…)" (Williams, 1946a, p. 18). Rather than a science of human beings for the sole sake of mere study and theoretical knowledge, Williams describes the potential and necessity for practical applications thereof on the final page of *The Human Frontier*: "*Humanity must understand itself*" (Williams, 1946a, p. 301). At the end of every chapter of *The Human Frontier*, the importance of the study of individuals is reiterated and impressed upon the reader. In this regard, *The Human Frontier*, serves as a call to arms for science and society to further the development of humanics. *The Human Frontier* aims to depict the necessity and potential for the study and practical application of knowledge of human individualities within the field of humanics. These potential applications are described in great detail within the confines of its 301 pages.

6.2 Audience

Preceded in its nature only by *What To Do About Vitamins* (Williams, 1945a), *The Human Frontier* is the second major publication by Roger Williams aimed at a broader audience (Williams, 1946a). The first indication toward this end can be found in the cover design of the hardback version as well as the book's title.[126] The stark contrast of dark green and yellow alongside white writing

---

125 See Sections 6.2. Audience, 9.3.1. Evidence, 9.3.3. Appeal, 10.1. Simple Yet Profound, 9.4. Biochemical Institute Studies, and 11.3. Individual Anatomies.
126 Fig. 5: Title Cover of the 1946 Hardback Version of The Human Frontier: A New Pathway for Science Toward a Better Understanding of Ourselves (Williams, 1946a) depicts the title cover of the 1946 hardback version of *The Human Frontier* (Williams, 1946a). An example for other, more restrained, publication covers is provided by Fig. 6 (Williams, 1935a).

renders the book prominent and eye catching in comparison to prior academic titles. Additionally, the title *The Human Frontier: A New Pathway for Science Toward a Better Understanding of Ourselves* creates a more dramatic impression when compared with the modest titles of his earlier publications.[127] Here, the title implies something revolutionary and ground-breaking, the promise of the acquisition of new knowledge and understanding. Subject to frequent changes and deliberations, other early titles for *The Human* Frontier include "Exploring Ourselves. A Post-War Frontier" (Pauling, 1946; Williams, 1945 f, 1945c, 1946c, 1946d, 1946l), "A Way to the Fulfilment of the Human Spirit", "A Way to Human Happiness", "Science Can Break the Chains That Bind Us", "The Truth About Ourselves Would Set Us Free" (Williams, Undated), and "Man, Frustrated or Free?" (Williams, Undated). This deviation from previous naming schemes is noteworthy, because it further highlights the alternative nature of this publication.

---

127   *What To Do About Vitamins* also constitutes a slight deviation from the very accurate and colloquial naming approach of Williams' textbooks (Williams, 1945a). However, while its name poses the question of the uses of knowledge on vitamins and suggests a solution is provided within the publication, it lacks the dramatic aspect of *The Human Frontier* (Williams, 1946a).

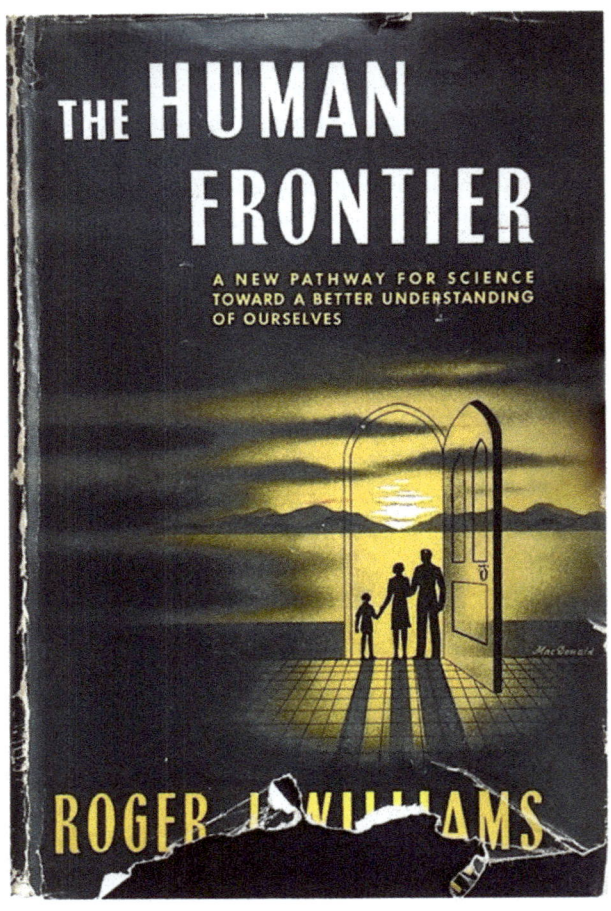

Fig. 5: Title Cover of the 1946 Hardback Version of The Human Frontier: A New Pathway for Science Toward a Better Understanding of Ourselves (Williams, 1946a)

Fig. 6: Title Cover of the 3rd Edition of *An Introduction to Organic Chemistry* (Williams, 1935a)

Furthermore, the publisher of *The Human Frontier* stands in stark contrast to all previously discussed publications. Williams' prior articles had all appeared in academic journals, while his (text)books were published by university and academic publishing houses. The choice of Harcourt, Brace and Company deviates from the aforementioned publishers in that they were much less specialised regarding their published works.[128] This is exemplified by their publication of the first edition of George Orwell's (1946) *Animal Farm* in the same year. The choice of publisher underlines the fact that this publication diverges from Williams' previous publications not only in its contents, but also pertaining to its desired audience. In correspondence discussing possible publishing houses, Williams highlights the importance of reaching the largest possible audience (Williams, 1945c, 1945 f). Following discussions with Williams, well-known writer and fellow scientist Paul de Kruif (1890–1971) plays an important role in *The Human Frontier*'s publication as an intermediary, facilitating the first contact between Williams and Harcourt, Brace and Company (de Kruif, 1945a, 1945b, 1945c; Newsem, 1946; Williams, 1946b, 1946e, 1946 f).[129]

---

128 The publishing house Harcourt, Brace and Company was purchased in 2007 and is now part of Houghton Mifflin Harcourt (Kojalo, 2007).

129 Paul de Kruif was an American microbiologist and well-known writer. For further information, see Krebs (1971).

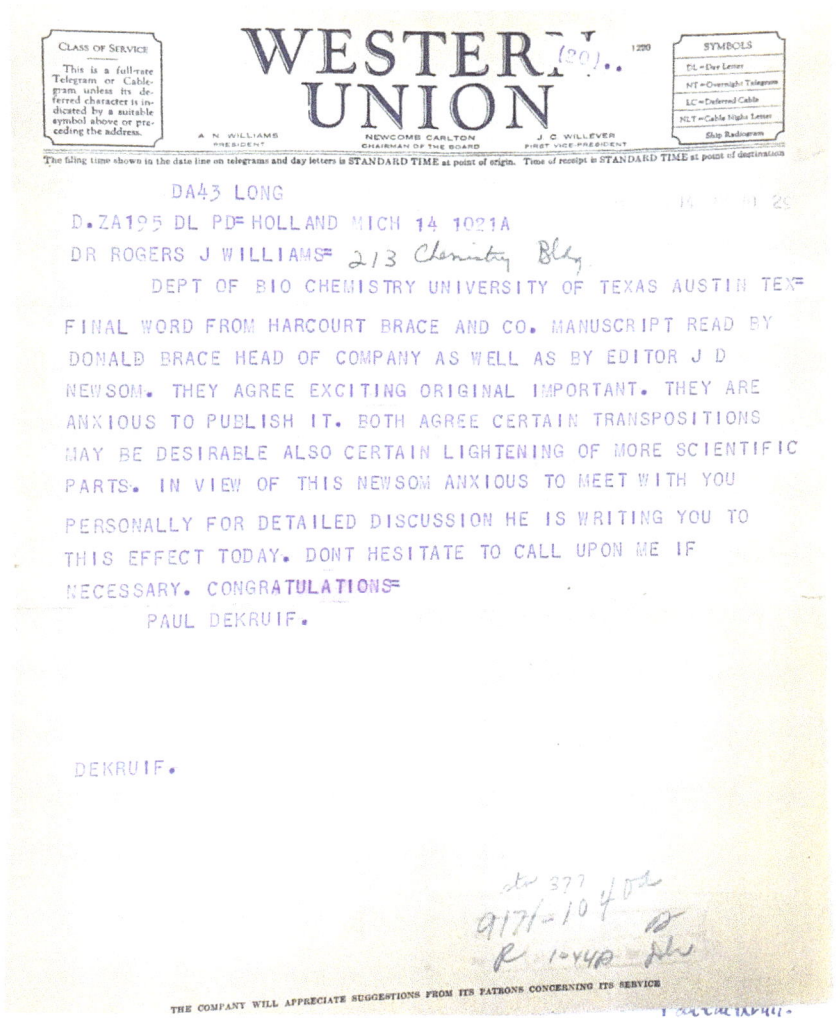

Fig. 7: Telegram by Paul De Kruif to Williams on 14.12.1945 confirming Harcourt, Brace and Co. is interested in publishing *The Human Frontier*; Dolph Briscoe Center for American History, The University of Texas at Austin, Roger Williams Papers, Box 88-087/26a, Folder: Correspondence Concerning Humanics Sept. 1945 – March 1946

HARCOURT, BRACE AND COMPANY, INC
PUBLISHERS
383 MADISON AVENUE, NEW YORK

January 9, 1946

Dear Paul:

Thank you! You have hit the right nail on the head with your customary vigour. A man-to-man talk, of course, is the only way to do a good job on the Williams manuscript. Don has read your note. He is as enthusiastic as I have been from the beginning about the book, make no mistake about it.

No other publisher need be approached. I have written Dr. Williams suggesting, first, that he come to New York at our expense, or second, if this is impossible, that I go to Austin and stay there until we have ironed out all minor wrinkles. Your marginal notes pleased me immensely, chiefly because they confirm my own reactions.

So I think I'll go to Texas in a few days. The book is far too important to let go and, speaking personally, I'd feel very badly if we didn't make every possible effort to help Williams put this important book into the best possible shape.

With love to you both.

Yours,

Jack
John D. Newsom

Dr. Paul De Kruif
Wake Robin
Holland, Michigan

JDN:BP

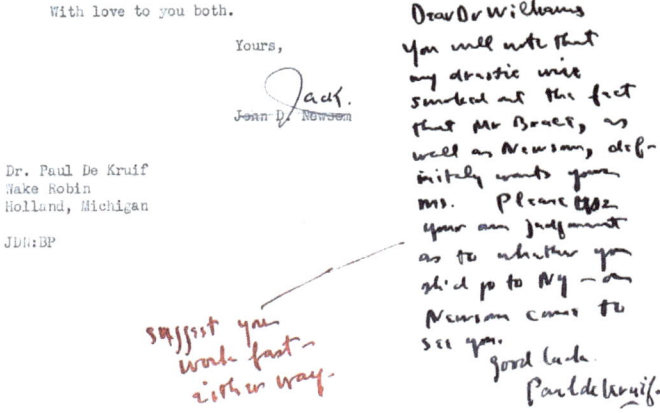

Fig. 8: Letter by John D. Newsem to Paul De Kruif on 09.01.1946 confirming that Harcourt, Brace and Co. is interested in publishing *The Human Frontier*; Dolph Briscoe Center for American History, The University of Texas at Austin, Roger Williams Papers, Box 88-087/26a, Folder: Correspondence Concerning Humanics Sept. 1945 – March 1946

During discussions pertaining to the final choice of publishing house, Williams writes an article entitled "Developing a Crucial Applied Science: Humanics". This article is later revised and appears in *The Scientific Monthly* as "Humanics: a crucial need", following a rejection to publish by *Science* (Williams, 1946g, 1947b). Sending pre-print copies to colleagues of all faculties at The University of Texas in 1946, Williams asks for opinions and interest in promoting the publication of *The Human Frontier* later that year (Williams, 1946h). Describing how his theories may have far-reaching consequences for all university faculties, a complimentary manuscript of *The Human Frontier* is offered to any interested party. Receiving overwhelmingly positive feedback, Williams is buoyed to continue pursuing the most widespread form of publication possible (Williams, 1946g). A further offer by a university publishing house is ultimately declined, as the possible outreach does not appear large enough (Williams, 1945 f).

May 20, 1946

Dear Colleagues:

The enclosed article was written for publication in <u>Science</u> and is only partially self-explanatory. It is being sent to all members of the faculty holding professorial status because of the possibility that it may lead to new developments in our university work and because any developments along the line indicated will be of concern not to a single division or department but to all of us.

I feel that the matter is of sufficient importance so that I have taken a considerable amount of time during the past year and a half to write a book, of which the enclosed is an attempted condensation. The book, entitled <u>The Human Frontier</u>, is now revised and edited and in the printer's hands. It is due for fall publication by Harcourt, Brace and Company. Several copies of a preliminary manuscript have been circulating on the campus, and the effort has received a substantial amount of praise both locally and from prominent individuals elsewhere.

If this material sounds interesting to you, will you please check appropriately on the slip below and return it to my office? If you have other comments, please feel free to make them.

Cordially yours,

Roger J. Williams
Chemistry Building 213

✓ I am favorably impressed and would like to be kept informed regarding developments.

✓ I would like to be invited to any preliminary conferences dealing with the problem outlined.

✓ I would like, when a copy is available, to see the preliminary manuscript referred to.

Other comments:

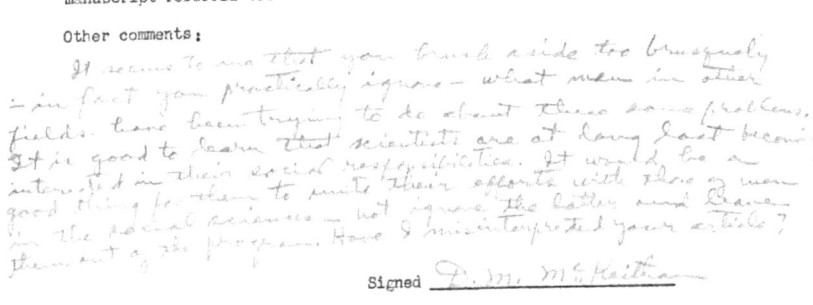

Fig. 9: Letter by Williams to D. M. McKeithan on 20.05.1946 asking for views on a manuscript of *The Human Frontier*; Dolph Briscoe Center for American History, The University of Texas at Austin, Roger Williams Papers, Box 88-087/26a, Folder: Humanics Article Faculty Reactions

The structure of *The Human Frontier*'s chapters further indicates its uniqueness when compared to Williams' previous publications. Each chapter begins with an opening quote referencing the individuality of humans, mostly by well-known personalities. Thematically linked to their corresponding chapters, these quotes serve as an introduction to the themes discussed therein. Ranging from authors and poets such as Alexander Pope (1688-1744),[130] George Eliot (1819-1880),[131] and Lord Byron (1788-1824),[132] the Greek philosopher Socrates (470 BCE-399 BCE),[133] American philosopher and theologian Henry Wieman (1884-1975),[134] physicist Albert Einstein (1879-1955),[135] and American president Theodore Roosevelt (1858-1919),[136] these quotes act as a demonstration of how widespread the concept of individualities existing in humans already were prior to the publication of *The Human Frontier* (Williams, 1946a, pp. 3, 20, 42, 81, 103, 185, 276). Additionally, they create an aura of self-evidence with regard to the contents of Williams' book. Readers may be much less likely to question statements when they are prefaced by a quote of a well-known and respected individual.

A further aspect in accordance with the aforementioned public nature of *The Human Frontier* is the language used regarding academic or scientific terms. In agreement with *What to Do About Vitamins*, *The Human Frontier* utilises terms such as "fuels" and "burning" when discussing human metabolism alongside the explanations of the term "enzyme" as a biochemical lubricant (Williams, 1946a, p. 21).[137] The changes made to the term "humanics" in the process of writing *The Human Frontier* also exhibit the importance of comprehensibility and general acceptance of the work as a whole. Originally christened "humanology" by Williams (1945g), he opts to change the term used to describe his new branch of science due to widespread critique from colleagues and confidants (Weaver, 1945, 1946). These felt the theories may be less readily accepted if the term "humanology" was used, as it lacked any familiarity (Williams, 1946b). Unhappy with the term even in first manuscripts (Williams, 1945g), and later describing it as "euphonius" and "inelegant" (Williams, 1946c), "humanology" is replaced by "humanics", a term adopted from a popular dictionary (Williams, 1946i, 1947c).

---

130 For further information and in-depth reading, see Butt (2023).
131 For further information and sources for in-depth reading, see Haight (2023).
132 For further information and sources for in-depth reading, see Marchand (2023).
133 For further information and sources for in-depth reading, see Kraut (2022).
134 For further information and in-depth reading, see Wieman and Peden (2010).
135 For further information and sources for in-depth reading, see Kaku (2023).
136 For further information and sources for in-depth reading, see Cooper (2023).
137 See Section 5.4 *What to Do About Vitamins*.

Another aspect indicating the importance of public approval is the number of unpublished manuscripts dealing with the same general topic, as well as the wide variety of colleagues consulted during the writing process of *The Human Frontier*. The importance of wide acceptance of his theories leads Williams to write multiple outlines and calls upon the advice of a plethora of colleagues from a variety of disciplines, including a United States Senator, in order to gauge public interest (Pauling, 1946; Weaver, 1945, 1946; Williams, 1945c, 1945 f, 1945h, 1945i, 1946c, 1946d, 1946i, 1946h, 1946g, 1946l, 1946m).

Finally, the general nature of *The Human Frontier* is confirmed by the reviews posted after its publication and the publication of abridged summaries of the book in non-scientific magazines. Reviews in the *New York Times* (Garside, 1947), *Marriage and Family Living* (National Council on Family Relations, 1948, p. 78), and *The Globe and Mail: Canada's National Newspaper* (Laurence, 1948) indicate its scope. It is further described as "appealing to the layman" (Glass, 1947, p. 176) and "in language addressed definitely to the lay reader" (The Human Frontier, 1947). A final review confirms that "the general public, for whom he writes" is the clear addressee (The Editors, 1949, p. 158). Condensed forms of *The Human Frontier* are presented in multiple magazines, including *The Reader's Digest* (Williams, 1948a),[138] *American Scientist* (Williams, 1947d),[139] and *THINK* (Williams, 1948b).[140] This widespread form of promotion further indicates a broader target audience.

Noted Cold War historian Lawrence Badash (1936–2010) offers a possible explanation for Williams' newly won fervour regarding public appreciation and

---

138   *Reader's Digest* describes its mission as follows: "Reader's Digest shares trusted advice and stories to help you and your family enjoy healthy, wealthy, and wise lives. We get to the heart of the matter and keep it simple, informative, and fun" (Reader's Digest, 2023). The family Magazine was first published in 1922 and sold more than one million copies every month at the time of Williams' article. It appeared in the United States, Denmark, the British Isles, Cuba, Brazil, and Sweden between August and July of 1948. For further information, see Reader's Digest (2023).

139   *American Scientist* describes itself as an "illustrated bi-monthly publication about science, engineering, and technology" on its website and has been published since 1913 (American Scientist, 2023). It is a well-known popular science publication. For further information, see American Scientist (2023).

140   THINK Magazine was published by the International Business Machines Corporation (IBM) and distributed free of charge as of June 1935. Its purpose was to "help to develop in its readers a more conscious and active impulse to *think*" (Watson, 1935, p. 3).

application of his theories from 1945 onwards.[141] American scientists, largely unknown to the public and media before the Second World War, started to gain substantial prominence in its wake (Badash, 2000, p. 55). The increase in medical and physical knowledge of the world, coupled with the relevance of this knowledge to the individual and society as a whole, meant scientists and their research increasingly appeared in popular media following the end of WWII. The success of The Manhattan Project offering further indication of the benefit of federal funding of scientific endeavours with "focus upon large problems" (Badash, 2000, p. 78),[142] the presentation of humanics as just such a "large problem" is an attractive alternative to the discreet and unobtrusive mode of purely academic publication Williams prefers prior to the war.[143]

As demonstrated by the choice of name, cover design, publisher, overall structure, terminology, reviews, and abridged publications, the target audience of *The Human Frontier* is the general public, rather than a purely scientific or academic community. In addition, the examples of individuality presented in *The Human Frontier* and *Biochemical Individuality* contain multiple overlapping elements.[144] This publication is, therefore, Williams' first clear appeal to the general public with regard to individualities of human behaviour and metabolism and, as such, presents a vital steppingstone within the development of BI.

However, the generalist nature of *The Human Frontier* does not mean its focus lies entirely outside of the academic world. As described above, an important aspect of *The Human Frontier* is its appeal for a more in-depth study of individual human beings.[145] While society and governments play an important role in calling for and funding such research, the scientific community must have the leading role in the execution of such research. The book, therefore, is an additional

---

141 Lawrence Badash was a Yale and Cambridge University educated historian and a professor of history at the University of California at Santa Barbara. His research focus was 20th century history, with a particular focus on physics and scientific history (Badash, 2000, pp. 79–80). For further information, see his published works and the published memoirs following his death.
142 The Manhattan Project is the codename for an American project in World War Two, undertaken in collaboration with the United Kingdom and Canada, with the aim of creating the world's first atomic weapons. For further information, see Rhodes (2012) or Kelly and Rhodes (2009).
143 See Section 7.5 Anti-Communist Sentiment.
144 See Tab. 1: Comparison Between Overlapping Citations in *The Human Frontier* and *Biochemical Individuality*.
145 See Section 6.1. "Humanity Must Understand Itself".

call to Williams' fellow scientists; it follows a similar goal of convincing the scientific community of humanics' importance. Furthermore, *The Human Frontier* acts as an appeal specifically to the medical community. Williams draws attention to the fact that, "even graduate medical students are usually not regarded as well-equipped for research" (Williams, 1946a, p. 229), and that further knowledge on the individual could help "medical treatment (…) meet more effectively the needs of individual patients" (Williams, 1946a, p. 231). *The Human Frontier*, therefore additionally acts as a plea for further education of medical students in order to make the individual study of human beings more effective and improve patient care. This could, according to Williams, reduce the number of individuals "beyond professional help" as well as situations "in which physicians admit that they are baffled and unable to be of assistance" (Williams, 1946a, p. 238). Published by a well-known publishing house and aimed at the broader public, the probability of a diverse group of scientists hearing of and reading *The Human Frontier* is increased. As Williams' concept of humanics relies on the joint research of multiple faculties, the publication of such an appeal by a specialised journal or publishing corporation would have been of little use in the actual recruitment of likeminded scientists or the persuading of others.

This two-sided nature of *The Human Frontier*'s appeal, the societal outcry and consequent bid for intensified scientific research on individuals, is reflected in its chapters. While topics relevant to society as a whole (marriage, employment, leadership, international relations, religion) are addressed, scientifically significant aspects (physiology, medicine and medical research, psychology, the senses, metabolism) are presented alongside these as equally relevant (Williams, 1946a, Table of Contents). The encouraging tone and eye-catching design of the book are aimed to inspire the general American public and scientists alike. *The Human Frontier: A New Pathway for Science Toward a Better Understanding of Ourselves* aims itself at a broader audience than all previously discussed works.

Tab. 1: Comparison Between Overlapping Citations in *The Human Frontier* and *Biochemical Individuality*

Citation	Page(s) of Citation in *The Human Frontier*	Page(s) of Citation in *Biochemical Individuality*
Heath, C. W. (1945). What people are; a study of normal young men. Harvard University Press.	11	8, 35, 52, 121
Loeb L. The Biological Basis of Individuality. 1st ed. Springfield, Illinois: Charles C Thomas, 1945.	26	132

Tab. 1: Continued

Citation	Page(s) of Citation in *The Human Frontier*	Page(s) of Citation in *Biochemical Individuality*
R. S. Banay, "Pathological Reaction to Alcohol," Quart. Jour. Stud, on Alcohol, 4, 580, 1944.	33	108
R. W. Engel, Proc. Soc. Expt. Biol. and Med., 52, 281–2, 1943.	38	156
R. F. Light and L. J. Cracas, Science, 87, 90, 1938.	38	149
C. J. Warder, H. C. Brown and S. Ross, Jour. Expt. Psych., 35, 57, 1945.	46	125
W. C. Halstead, Science, 101, 615, 1945 303.	46	124
A. F. Blakeslee, Science, 81, 504–7, 1935.	71	127
A. F. Blakeslee, Proc. Nat'l Acad. Sci. USA, 21 (2)78–83, 84–90, 1935.	71	127
A. F. Blakeslee, Jour. of Heredity, 23, 106, 1932.	74	130
E. A. Hines, Jr., and G. E. Brown, American Jour. Heart, *ii*, i, 1936.	83	125
N. Kleitman, Sleep and Wakefulness, Univ. of Chicago Press, 1939.	92	122
A. Grollman, Essentials of Endocrinology, J. B. Lippincott, 1941.	110	81
M. A. Goldzieher, The Endocrine Glands, p. 860, D. Appleton-Century Co., 1939.	113	90
A. T. Rasmussen, Amer. Jour. Anat., 42, i, 1928.	116	91, 92
A. T. Rasmussen, Endocrinology, 8, 509, 1924.	116	91, 92
A. T. Rasmussen, Endocrinology, 12, 129, 1928.	116	91, 92
Riddle O. Endocrines and Constitution in Doves and Pigeons. First Edition. Washington, D.C.: Carnegie Institute of Washington, 1947.	120	19
Advertisement appearing in Jour. Am. Med. Assoc., Oct. 20, 1945.	230	143
L. S. Goodman and A. Gilman, The Pharmacological Basis of Therapeutics, The Macmillan Co., 1941.	232	111
G. Draper, C. W. Dupertuis and J. L. Caughey, Jr., Human Constitution in Clinical Medicine, Paul B. Hoeber, Inc., New York, 1944.	233, 243	2, 190

(*continued*)

Tab. 1: Continued

Citation	Page(s) of Citation in *The Human Frontier*	Page(s) of Citation in *Biochemical Individuality*
S. H. Kraines, The Therapy of Psychoses and Neuroses, Lea and Febiger, Philadelphia, 1943.	235	201
E. S. Gordon, Chapter in Biological Action of the Vitamins, E. A. Evans, Editor, Univ. of Chicago Press, 1942.	247	154

6.3 Social Control

An aspect of *The Human Frontier: A New Pathway for Science Toward a Better Understanding of Ourselves* which necessitates discussion is the issue of social control. As touched upon in Section 6.1 "Humanity Must Understand Itself", Williams sees one benefit of the scientific study of humanics in the potential "improvement in social control" (Williams, 1946a, p. 5). Such study of individuality with the ultimate goal of improved social control raises thoughts akin to Lois Lowry's dystopian novel *The Giver*, in which every individual of a society is given the ideal medication for sensory deprivation, according to their individual make-up (Lowry, 1993). The term "social control" brings with it the negative connotation of totalitarian societies in which individuals are sorted and forced into labour in predetermined occupations according to their individual talents. Williams' use of this term may indicate such unappealing thoughts in connection with *The Human Frontier*. A historical, political, ethical, and philosophical discussion of the concept of "social control" would exceed the boundaries of this dissertation.[146] Conversely, this section rather seeks to clarify Williams' view of the possibilities for humanics regarding society, which differs from the dystopian ideas the term may elicit.

The following excerpts from *The Human Frontier* provide an example of individuality which humans have already been able to address for centuries. Williams' critique is based on the fact that human social interactions and policies do not address other similarities in the same way.

> There are no two pairs of feet in America that are precisely alike, toe prints and all, and offhand it might seem impossible to use large-scale methods and at the same time fit everyone with comfortable shoes. Actually it is not so difficult. The majority of people

---

146 For a definition and extended discussion hereof, see Horwitz (1990).

get along very satisfactorily if their shoes are approximately the right length and width and are built according to a standard last. Some feet require special attention for maximum comfort.

Our attempted social adjustments, however, are often so crude that they might be compared to furnishing an entire army with *average sized* shoes. For the purposes of calculating the amount of leather required to put shoes on an army it would be valuable to know the average size of the soldiers feet, but this information would be of no value in ordering the sizes to be made. An average-sized shoe would fit very few individuals. (Williams, 1946a, pp. 13–14)

(…)

A knowledge of how human beings range in variability in each of the human activities will make it possible to develop systems of classification which can be used by individuals as bases for their personal adaptation. Until the ranges of variability are known and until each individual is in a position to learn at least in a broad outline about his own attributes and aptitudes, there will continue to be an enormous number of misfits – cases where society is attempting to force average size shoes onto people whose feet are far from average. (Williams, 1946a, p. 165)

This excerpt demonstrates Williams' understanding of social control, and he later describes it as the best example of society caring for individuality known to him (Williams, 1950b, p. 271). The term does not propose the more effective control of human beings by others, but instead suggests the benefit of increased self-knowledge. Through approximated classification of "nearly alike" individuals, society could increase their effectiveness in choosing appropriate activities and occupations tailored to their personal preferences and capabilities (Williams, 1946a, p. 13). Williams postulates that "no matter what the activity is, it should be society's goal to provide for the variability which exists in that activity" (Williams, 1946a, p. 165). A situation in which, socially, humans are deemed average and therefore are forced to wear metaphorical "average shoes" cannot provide the necessary support or fit for the individual to prosper. The ultimate goal of Williams' concept of social control is, therefore, "social welfare" (Williams, 1946a, p. 13).

A perhaps more suitable term for this concept, employed later in *The Human Frontier*, is "social adjustment" (Williams, 1946a, p. 166). Alternatively, the term "social betterment" also describes Williams' concept of social control more accurately (Williams, 1946a, p. 125). Williams' usage of the term social control must, therefore, be viewed with relation to his own interpretation of the expression, as the use of other definitions could misconstrue a fundamental motivation behind *The Human Frontier*.

6.4 "Distinctive Metabolic Traits"[147]

Following the introductory remarks of *The Human Frontier*, Williams begins to argue his case for the development of humanics by presenting a rationale for individualities in human behaviour. Portraying evidence toward fundamental differences in metabolic patterns as the root of the aforementioned individualities, Williams presents ideas akin to those offered in *Biochemical Individuality: The Basis for the Genetotrophic Concept*. This section is largely dedicated to the analysis of the second chapter of *The Human Frontier*, "Fundamental Metabolism as It Is Related to Character Traits" and the sources cited therein (Williams, 1946a, Chap. II). Collectively, they provide an indication of further inspiration for, and influence on, the manifestation of BI within the pages of *The Human Frontier: A New Pathway for Science Toward a Better Understanding of Ourselves*.

The second chapter of *The Human Frontier* offers elucidations of scientific studies providing evidence for biochemical and genetic idiosyncrasies of human individuals. In part, these explanations involve extrapolation from animal research, though Williams (1946a, p. 21) postulates that "metabolism in man and higher animals has many resemblances" and "the general features of the over-all process of metabolism are essentially the same for men and for higher animals" (Williams, 1946a, p. 20). This constitutes a paradigm-shift when related to Williams' principles of biochemical research.[148] While this basic scientific research of biochemical molecules and processes is best studied using less complex life forms, research of BI is primarily based on the study of higher life forms.[149] Williams' later research of BI is similarly based on the study of rats and other vertebrates.[150]

Having clarified the basis of the research presented, Williams presents the first evidence of the cause for the uniqueness of every human being.

---

147 (Williams, 1946a, p. 25). To improve the clarity of this section-title and avoid confusion, the citation for this quote is presented in this footnote, deviating from the citation style used in the rest of this dissertation. The capitalisation of this quote has been altered to fit the style of section-titles of this dissertation. The original quote reads "distinctive metabolic traits".
148 For further reading on the topic of paradigm-shifts and scientific progress, see Kuhn and Hacking (2012).
149 See Section 4.2. "The Vitamin Requirement of Yeast".
150 See Chapters 7. The Scientific Study of Individuals and Alcoholism, 8. Genetotrophic Disease, and 9. Human Individuality.

We are beginning to have definite and striking proof that this heritable machinery includes the specialized enzymes or the ability to produce them as clarified by Beadle and his co-workers, and that it is for this reason that our metabolism follows a pattern derived from those of our forebears. Two animals which inherit identical anatomical structures have the machinery for carrying on metabolism in exactly the same way, but if there are differences in their anatomies (including microscopic and sub-microscopic details) then their metabolism will show corresponding variations. (Williams, 1946a, p. 21)

Postulating genetic variation as the basis for biochemical discrepancies, Williams presents the research of geneticist G. W. Beadle (1903–1989) on *Biochemical Genetics* (Beadle, 1945).[151] In this 96-page manuscript, published a year prior to the release of *The Human Frontier*, Beadle reiterates the "one gene – one enzyme" hypothesis posed with E. L. Tatum (1909–1975) in 1941 (Beadle and Tatum, 1941, p. 499). Thereby, Beadle suggests that "genes directly determine enzyme specificities and thereby control in a primary way enzymatic syntheses and other chemical reactions in the organism" and that "a given enzyme will usually have its final specificity set by one and only one gene" (Beadle, 1945, p. 19). Extrapolating from the knowledge gained through Beadle's work, Williams presents his concept on which BI is later based.[152]

> (…) *each human being has a metabolic pattern which differs in some respects from that of all his fellows.* This fact is important for it is probably upon these fundamental metabolic differences that our observable individual differences rest. (Williams, 1946a, pp. 21–22)

This hypothesis constitutes the central presumption on which *The Human Frontier* is based, an aspect highlighted by Williams' use of emphasis. Additionally, this idea is mirrored by the second chapter of *Biochemical Individuality: The Basis for the Genetotrophic Concept*, "Genetic Basis of Biochemical Individuality"

---

151 G. W. Beadle is best known for his work on the "one gene – one enzyme" hypothesis with E. L. Tatum, and later became a later Nobel-laureate (Beadle, 1958, p. 592). For further information on his life and work, see Srb (1990).

152 Williams qualifies the importance of Beadle and Tatum's work in *Biochemical Individuality: The Basis for the Genetotrophic Concept*: "Our discussion is by no means dependent upon the acceptance of any simple 1-gene-1-enzyme relationship, but it does rest upon the widely substantiated principle that the potentiality possessed by organisms for carrying out any and every chemical reaction arises from inheritance and intervening mutations" (Williams, 1956a, p. 10). Williams further describes and cites the work of Beadle and Tatum (1941) with regard to *Neurospora* as a basis for the concept of partial genetic blocks, which in turn plays an important role in the effectiveness of enzyme systems relevant to BI (Williams, 1956a, p. 11).

(Williams, 1956a, Chap. II). The importance of Beadle's work to Williams' theory is highlighted by its reiteration in many following publications (Williams, 1948c, p. 50; Williams and Beerstecher Jr., 1948a, p. 412; Williams et al., 1949a, p. 275; Williams, 1956a, p. 10).

In close relation to the work of Beadle on *Biochemical Genetics*, Archibald Garrod's (1857–1936) work on *Inborn Errors of Metabolism* (second, revised edition) is of similar relevance to the ideas posed in *The Human Frontier* (Garrod, 1923). Quoted both by *Biochemical Genetics* and *The Human Frontier* (Beadle, 1945, p. 90; Williams, 1946a, p. 303), Garrod's research has been widely considered as central to the development of personalised medicine and the idea of individuality in humans it is based on. Williams later quotes Garrod's original work from 1902 in *Biochemical Individuality* (Williams, 1956a, p. 97), in which Garrod proposes a concept of human individuality sharing similarities with BI. In the 216-page manuscript from 1923, Garrod proposes that metabolic variations are at the root of differences between and within species (Garrod, 1923, p. 2). Elucidated by the differences we see in human beings (skin and hair colour, etc.) and describing "idiosyncrasies with regard to drugs and articles of food" (Garrod, 1923, p. 3), further coinciding aspects of Garrod's and Williams' theories become evident.[153] Though Garrod goes into much further chemical detail, describing explicit examples of differing metabolism pertaining to specific chemical compounds, many aspects of his work are mirrored in *The Human Frontier* and BI. In his first publication on chemical individuality, Garrod proposes a theory very similar to that of BI and the thoughts presented in *The Human Frontier*. In Garrod's statement that "just as no two individuals of a species are absolutely identical in bodily structure neither are their chemical processes carried out on exactly the same lines" (Garrod, 1902, p. 281), the similarities between the *The Human Frontier* and "The Incidence of Alkaptonuria: A Study in Chemical Individuality", and therefore relevance of Garrod's work to BI, become clear.[154]

Though they are published 23 years apart, and therefore under drastically different conditions regarding the knowledge of biochemistry available, some of Garrod's core assumptions are mirrored in Williams' work.[155] Suggesting that, "to

---

153   See Section 6.5. "Individuals Vary Greatly".
154   The work of Archibald Garrod has been intensively researched and discussed in academic literature and will therefore not be further expanded on at this point. For further information, see Perlman and Govindaraju (2016).
155   Garrod describes the scientific context of his publication as follows: "The conception of metabolism in block is giving place to that of metabolism in compartments. The view is daily gaining ground that each successive step in the building up and

the students of heredity the inborn errors of metabolism offer a promising field of investigation" (Garrod, 1923, p. 18), the idea of hereditary aspects being relevant to metabolic diseases is presented. From this, the concept of a heredity of general metabolism can be extrapolated, two aspects later relevant to Williams' "Genetotrophic Principle".[156] Furthermore, Garrod elaborates on multiple heritable metabolic diseases including albinism, alkaptonuria, cystinuria, and others (Garrod, 1923, Table of Contents). These diseases are used as examples of individuality in *The Human Frontier* (Williams, 1946a, p. 24). This overlap between the topics covered by Garrod and Williams indicates the relevance of Garrod's work for the development of BI in *The Human Frontier*.

A further scientist's work to be considered in the second chapter of *The Human Frontier* is that of pathologist Leo Loeb (1869–1959).[157] Author of *The Biological Basis of Individuality*, his work on the transplantation of tissues is considered in its relevance to individualities in humans (Loeb, 1945). It is noteworthy that the previously mentioned 711-page book, which cites and discusses many different aspects of individuality, is not cited in *Biochemical Individuality*. Furthermore, it is not cited in other sections of *The Human Frontier*, though topics discussed in Loeb's work, such as blood groups (Loeb, 1945, pp. 478–497) and psychical-social individuality (Loeb, 1945, pt. VIII), are also considered in *The Human Frontier* (Williams, 1946a, pp. 26–27, 125–161).[158] While not relevant in other aspects of Williams' publications, the citation of *The Biological Basis of Individuality* (Loeb, 1945) is noteworthy because it provides a comprehensive summary of other research on individuality prior to the publication of *The Human Frontier*. Additionally, this work is later reconsidered in *Biochemical*

---

     breaking down, not merely of proteins, carbohydrates and fats in general, but even of individual fractions of proteins and of individual sugars, is the work of special enzymes set apart for each particular purpose" (Garrod, 1923, p. 5). During the 23 years between the publications of *Inborn Errors of Metabolism* (Garrod, 1923) and *The Human Frontier*, the research within this scientific field had progressed in such a fashion that Williams speaks of these as self-evident.

156 See Chapter 8. Genetotrophic Disease.
157 Leo Loeb was a pathologist who published extensively on individuality. For further information on his life and work, see Goodpasture (1961).
158 The section on blood groups in *The Human Frontier* is discussed directly following the citation of Loeb's research with regard to transplants and is not cited separately (Williams, 1946a, p. 26). It is therefore not entirely clear whether the information on blood groups stems from Loeb's *The Biological Basis of Individuality* or another source. As Williams tends to cite the source of information at the end of his descriptions, and no source is offered here, it may stem from another source.

Individuality (Williams, 1956a, p. 132). Described as "monumental work", Loeb's research is mentioned, though not further discussed, "because the evidence is voluminous and has been treated fully" (Williams, 1956a, p. 132). These quotes could be a further indication for the reasoning behind the sparing analysis of Loeb's research in *The Human Frontier*. Nonetheless, the thoughts presented therein, in part explicitly cited by Williams at later points (Williams, 1953a, p. 17, 1954d, p. 795), will have shaped Williams' thoughts and ideas within the realm of BI.

The final example of influential research of individuality discussed in both *The Human Frontier* and *Biochemical Individuality* is the issue of disease susceptibility. In this regard, Williams cites the research of George Draper (1880–1959) and his colleagues at the *Constitutional Clinic of the College of Physicians and Surgeons* at Columbia University.[159] Their work on the susceptibility of individuals to certain diseases, presented in *Human Constitution in Clinical Medicine* (Draper et al., 1944), is of substantial relevance to Williams. Draper is named alongside famous and influential promoters of individuality, including Hippocrates (460 BCE–375 BCE) and Galen (129–216),[160] [161] in the introduction of *Biochemical Individuality* (Williams, 1956a, p. 2). Draper's research on constitutional medicine, in which the definition of constitution as the "aggregate of hereditarial characters, influenced more or less by environment, which determines the individual's reaction (…) to the stress of the environment" (Draper, 1925, p. 526) is proposed, is described at length in *The Human Frontier*. Draper's definition of constitution largely overlaps with Williams' concepts regarding vitamin deficiencies in "normal" diets discussed elsewhere.[162] His focus in *The Human Frontier* lies on Draper's research of a possible connection between certain phenotypes (such as eye-lash length and tooth gap size) and disease

---

159 George Draper was a medical doctor and is best known as the personal physician of President Franklin D. Roosevelt (1882–1945) (Hart, 2014). He and his colleagues are seen as the founders of the movement promoting "constitutional medicine". For further information on his life and work, see Comfort (2012).

160 Hippocrates was an ancient Greek physician and is often titled as the "father of medicine". For further information on his life and work, see Smith (2022).

161 Galen, an ancient Greek physician and philosopher, produced many influential works on medical theory. For further information on his life and work, see Nutton (2023).

162 See Sections 5.3. Practical Applications of Vitamin Research, 6.5. "Individuals Vary Greatly", 8.1. The Metabolic Individualities of Rats, 9.3. *Nutrition and Alcoholism*, and 11.1. Genetotrophic Supplementation.

susceptibility to poliomyelitis (Williams, 1946a, pp. 240–241). The idea of individual susceptibility to a number of diseases, especially those of a nutritional nature and alcoholism, become a central point of study in Williams' later career and are central to the Genetotrophic Concept.[163]

When regarding the works of the aforementioned scientists and their relevance to the ideas presented in *The Human Frontier*, the importance of Williams' own research must not be forgotten. Though he only cites one prior work of his own ("'Taste Deficiency' for Creatine"), the evidence of individuality discussed in previous sections is not diminished in its importance (Williams, 1946a, p. 304). A large portion of the examples of individuality presented in *The Human Frontier* are based on Williams' own experiences and observations.[164] While the evidence of Williams' inspiration up to the publication of *The Human Frontier* arises from within his own work, outside influences on his ideas become evident here for the first time.

As in other previously discussed instances, scientific and academic evidence of individuality is not the sole source of inspiration for the concepts presented in *The Human Frontier*.[165] Examples of individuality presented in culture and literature are regularly illustrated by Williams. One such example, which appears in multiple following publications, is "the old nursery rhyme about Jack Sprat and his wife" (Williams, 1946a, p. 22, 1947e, p. 574, 1951a, p. 18, 1951b, p. 19, 1953a, p. 95, 1956a, p. 160).[166] Through such examples, which are assumed to be "familiar", the previously discussed air of self-evidence is increased and the attention of the reader further captured (Williams, 1947e, p. 547).[167] A poem discussing similar topics appears in *Biochemical Individuality*.[168] The works of Beadle, Garrod, and Loeb provide information on the broader basis of, and influences on, the

---

163   See Chapter 7. The Scientific Study of Individuals and Alcoholism and Sections 8.1. The Metabolic Individualities of Rats, 8.3. Genetotrophic Promotion, 9.1. Biochemical Individuality V, 9.3. *Nutrition and Alcoholism*, 9.4. Biochemical Institute Studies, 11.1. Genetotrophic Supplementation, and 11.5. Normal Young Men.
164   See Section 6.5. "Individuals Vary Greatly".
165   See Chapters 3. A Career in Biochemistry and 4. Origins of Biochemical Individuality as a Concept.
166   Multiple versions of this poem exist, each with slight variations in their wording. A common version reads as follows: "Jack Sprat could eat no fat, His wife could eat no lean. And so between them both, you see, They licked the platter clean" (Opie and Opie, 1977).
167   See Section 6.2 Audience.
168   "It's a very odd thing –, As odd as can be –, That whatever Miss T. eats, Turns into Miss T" (Williams, 1956a, p. 168).

development of Roger Williams' concept of BI. With, in certain aspects, sizable commonalities between these publications and *The Human Frontier*, their influence on the development of BI is pronounced.

6.5 "Individuals Vary Greatly"[169]

Following the description of the external scientific basis on which Williams' theories are founded, the largest part of *The Human Frontier* provides examples of individuality from science and day-to-day life. Some examples of individuality presented in *The Human Frontier* are discussed in the following section. For the sake of clarity and brevity, not all examples of individuality presented in *The Human Frontier* are described here. First, those topics discussed in previous sections of this dissertation will shortly be addressed, followed by a few representative examples reoccurring in *Biochemical Individuality*. An in-depth discussion of the sociological theories and elaborations is not included, as they have been addressed in Section 6.1 "Humanity Must Understand Itself".

Having amassed a considerable amount of evidence of human individualities, Williams presents multiple examples discussed in previous publications. With regard to the individual scents of human beings, the identical example of scent discrimination in bloodhounds is presented, and is explored only marginally further (Williams, 1946a, p. 25).[170] Similarly, the variation in the detection of skunk odour by humans is described as previously (Williams, 1946a, p. 75). The aspect of creatine taste deficiency is additionally elucidated in the previously presented fashion, though the social aspects thereof are also shortly addressed (Williams, 1946a, p. 70). Williams thereafter describes research on the taste deficiency of phenyl thiocarbamide[171], the discussion of which reappears in later publications and in *Biochemical Individuality* (Williams, 1947e, p. 575, 1953a, p. 27, 1956a, p. 127).

In contrast, the issue of adverse drug reactions is discussed at length. With morphine and novocaine recurring (Williams, 1946a, p. 29),[172] Williams (1946a,

---

169   (Williams, 1946a, pp. 54–55) To improve the clarity of this section-title and avoid confusion, the citation for this quote is presented in this footnote, deviating from the citation style used in the rest of this dissertation. The capitalisation of this quote has been altered to fit the style of section-titles of this dissertation. The original quote reads "individuals vary greatly".
170   See Section 5.3. Practical Applications of Vitamin Research.
171   "Phenyl Thiocarbamide" is also referred to as "PTC" within this dissertation.
172   See Sections 4.3. Adverse Drug Reaction and 5.3. Practical Applications of Vitamin Research.

pp. 27–35) extends his list of examples significantly, describing similar effects for barbiturates, sex hormones, adrenaline, aspirin, digitalis, and penicillin, amongst others, as well as the idea of variability in susceptibility to caffeine, nicotine, and alcohol. While the individualities regarding caffeine and nicotine neither feature in *Biochemical Individuality: The Basis for the Genetotrophic Concept* nor Williams' later research, the issue of alcoholism becomes a topic of intense interest and study at DBUT in later years.[173] Aspects such as pathological intoxication (p. 33), varying signs of intoxication at identical blood alcohol levels (p. 32), the absence of moderate drinking in sober alcoholics (p. 34), and racial differences regarding alcohol addiction (p. 34) are discussed (Williams, 1946a). These aspects all reappear in following publications on alcoholism. In *The Human Frontier*, this research is proposed by Williams for the first time; the first publication from DBUT on this topic is released the following year (Williams, 1947e).[174]

Finally, the uncertainty around the exact vitamin requirements of human beings, and the supposed interindividual differences with regard thereto, constitutes the last example reflecting prior publications by Williams. Previously discussed theories and research are presented once more, with the new aspect of illnesses arising from a lack of certain vitamins being newly introduced (Williams, 1946a, p. 39).[175] Williams underlines the fact that the cited research projects "were not planned with this thought in mind", yet confirms they present to be "wholly in line with the idea that individuals may have distinctly different requirements" (Williams, 1946a, p. 39). Additionally, the therapeutic potential of vitamin supplementation appears with the elucidations of vitamin deficiency. These statements support the hypothesis of *The Human Frontier*'s raison d'être discussed previously.[176] The issue of individual vitamin requirements remains relevant to Williams throughout his career and thus reappears in *Biochemical Individuality* as well as his other works on nutrition.[177]

These examples further demonstrate the nature of the development of BI as an accumulation and expansion of knowledge over decades of research. The continued reoccurrence of similar or expanded examples of individuality show the process of the development of BI as a concept. Twenty-three of the sources cited in *The Human Frontier* also appear in *Biochemical Individuality: The Basis for the*

---

173   See Sections 7.2. "The Etiology of Alcoholism" and 7.3. "Alcoholics and Metabolism".
174   See Section 7.2. "The Etiology of Alcoholism".
175   See Section 5.4. *What to Do About Vitamins*.
176   See Section 6.1. "Humanity Must Understand Itself".
177   For full lists of Williams' publications, see Davis (2003b, 2003e).

Genetotrophic Concept (Williams, 1946a, pp. 303–308, 1956a, pp. 6–7, 17, 45–46, 66–68, 78–79, 95–96, 104–105, 117–118, 132–134, 162–165, 176, 195–196, 208–209).[178] These range from psychological and endocrinological research to studies on sleep behaviour and advertisements in medical journals. This considerable overlap is a further indication of the importance of *The Human Frontier* within the development of BI. The fact that approximately 1/5 of those sources relevant in 1946 reappear in *Biochemical Individuality* shows that the research and information provided by *The Human Frontier* serve as a basis for further work on BI.[179]

While large overlaps between the two books exist, discrepancies between several citations in *Biochemical Individuality* and *The Human Frontier* indicate the differing states of research regarding BI at the respective times of publication.[180] Prior to 1946, Williams had not released a publication with the explicit object

---

178 Twenty-two of these sources appear in *The Human Frontier*'s bibliography (Williams, 1946a, pp. 303–308). The final overlapping source is referenced, but never formally cited. On page 120, the research of a scientist "Riddle" on doves with regard to prolactin and size of thyroid glands is described (Williams, 1946a, p. 120). *Biochemical Individuality* cites exactly such a work in Chapter three under "²Oscar Riddle, *Endocrines and Constitution in Doves and Pigeons*, Carnegie Institution of Washington, Washington, D.C., 1947." (Williams, 1956a, p. 46). The exact topics referenced in *The Human Frontier* are also discussed in Chapters 3 and 4 of *Endocrines and Constitution in Doves and Pigeons*, even though Riddle's work is published after the release of *The Human Frontier* (Riddle, 1947). Lacking proper citation, it is impossible to clearly denote the source of Williams' information in *The Human Frontier*. However, the assumption can be made that the information available to Williams at the time of writing of *The Human Frontier* is also represented in *Endocrines and Constitution in Doves and Pigeons*. As Riddle's book had been completed and approved for publishing in 1945 (Riddle, 1947, p. iv), there is the possibility that Williams had access to the results and information before the actual publication of the research in book form. A close professional collaboration between the two is additionally implied by Riddle's inclusion in Williams' acknowledgement and thanks in the preface to *Biochemical Individuality* (Williams, 1956a, p. xi). Therefore, this reference to Riddle will be treated as a twenty-third overlapping citation between *The Human Frontier* and *Biochemical Individuality*. For the full list of sources, see Tab. 1: Comparison Between Overlapping Citations in *The Human Frontier* and *Biochemical Individuality*.

179 $(23/108) * 100 = 21.3\ \%$.

180 *The Human Frontier* cites 107 sources (Williams, 1946a, Bibliography) in its entirety, while BI cites 478 sources (Williams, 1956a, pp. 7, 17, 46, 68, 79, 96, 105, 118, 134, 165, 176, 196, 209).

of describing individualities in human beings. *The Human Frontier*, therefore, serves as a trailblazer in this regard. Following the publication of *The Human Frontier*, Williams releases a multitude of publications on this topic which further develop his theory, finally culminating in the publication of *Biochemical Individuality*. Williams is so convinced of the truth of his theory at this point that he funds the initiation of a special nursery in Austin, providing that it follows the goal of appreciating and learning about the aptitudes and characteristics of each individual child (Williams, 1946j).

Contrary to the style in which information is presented in *Biochemical Individuality*, *The Human Frontier* is much less scientific in its approach.[181] Williams' previous scientific papers and publications are filled with numerous citations, while *The Human Frontier* places less focus on these. While the Bibliography of *The Human Frontier* contains 107 citations, many more examples of individuality are presented in an offhand, anecdotal fashion. Including an example discussed in a previous section of this dissertation,[182] most of the descriptions of individuality presented in *The Human Frontier* are styled in the form of personal observations and remain uncited. This pattern is in accordance with the previously described assertion that *The Human Frontier* is less of a scientific publication and aimed at the general public.[183]

An example of such anecdotal evidence is Williams' account of a golf game he had previously played with officers of the United States Army.

> One of them, a big husky, was having considerable difficulty. Finally, after missing a shot, he remarked, "That wind makes the tears come to my eyes so I can't see the ball," which caused his partner, whose eyes were unaffected, to rib him cheerfully about the quality of his alibi. The facts are of course that the tear glands of different individuals vary greatly in their tendency toward activity. I watched the husky soldier thereafter and it was plain that the watering of his eyes was giving him a lot of trouble, while his partner was perfectly dry-eyed. We learn from childhood to think of ourselves as normal and to judge others accordingly. Consciously or unconsciously we say to ourselves, "This other fellow is just another such as myself. If he appears to be bothered by wind or glare when I am not, it must be a pretense – he is letting his imagination play tricks on him". (Williams, 1946a, pp. 54–55)

This excerpt is exemplary for the structure used throughout *The Human Frontier* to impress the importance of individual differences upon the reader. Personal accounts with which many readers will be able to relate precede descriptions of

---

181 See Section 12.1. Evidence.
182 See Section 4.1. Youthful Observations.
183 See Section 6.2. Audience.

typical social phenomena. The chapters of *The Human Frontier* are loose collections of thematically linked anecdotes, intermittently bolstered with scientific evidence and research. This evidence includes three diagrams and three tables (Williams, 1946a, pp. 48, 57, 64, 89, 100, 116). In contrast, *Biochemical Individuality* contains far higher numbers of diagrams and tables, which additionally play a more prominent role.[184] *Biochemical Individuality* is entirely different in its focus. When regarding the list of overlapping citations between the two books, it becomes clear that all of these are scientific articles or books, none are dedicated to sociological topics.

An example of such an overlapping citation in which the source is given considerable attention is the research presented by Clark W. Heath (1900–1986) and his colleagues in "What People Are: A Study of Normal Young Men" (Heath et al., 1945).[185] Though the differing styles of citation in *Biochemical Individuality* and *The Human Frontier* could be misunderstood to convey otherwise, both publications reference this study of "normal young men" as proof of the individualities of composition in humans.[186] *The Human Frontier* devotes one short

---

184   See Section 12.1. Evidence.
185   Clark W. Heath was a medical researcher based at Harvard Medical School in Massachusetts. He was the principal investigator of Harvard University's Grant Study, the summary of which is published as "What People Are: A Study of Normal Young Men". For further information, see Francis A. Countway Library of Medicine, Center for the History of Medicine. (2023).
186   *The Human Frontier: A New Pathway for Science Toward a Better Understanding of Ourselves* cites the article as "C. W. Heath, et al., *Normal Young Men*, Harvard Univ. Press, 1945" (Williams, 1946a, p. 303), while *Biochemical Individuality: The Basis for the Genetotrophic Concept* presents two forms of citation: "Clark W. Heath, *et al.*, What People Are, *Harvard University Press*, Cambridge, Mass. 1946" (Williams, 1956a, p. 66) and "Clark W. Heath, et al., *What People Are*, Harvard University Press, Cambridge, Mass., 1945." (Williams, 1956a, pp. 46, 132). No publications under these exact titles have been found at the time of writing. The only publication by Heath and his colleagues from this time fitting the requirements is titled "What People Are: A Study of Normal Young Men". From this, it can be extrapolated that all three citations fail to depict the full title of the publication, each citing only half of its heading. The discrepancy in citation regarding the year of publication may merely be an error in typing. It could additionally be explained by the fact that the second print of "What People are: A Study of Normal Young Men" appeared in 1946, while the first printings were available for purchase in 1945 (Heath et al., 1945, 1946). As Heath is cited with the correct year on other occasions in *Biochemical Individuality*, a typing error seems most likely. Both are not to be mistaken with Williams' own articles "Metabolic Peculiarities in Normal Young Men as Revealed by Repeated

paragraph to its examination, in contrast *Biochemical Individuality* quotes "What People Are: A Study of Normal Young Men" in four instances (Williams, 1946a, p. 11, 1956a, pp. 8, 35, 52, 121). It is worth noting that different sections of Heath et al.'s publication with correspondingly differing themes are cited and discussed in each instance. The underlying importance of these overlapping citations is, however, not dependent on the reuse of concrete examples. *The Human Frontier* represents a first attempt to provide evidence for BI in a public statement and acts as a forerunner to *Biochemical Individuality*. Therefore, the reappreciation of identical sources is of importance, though the concrete examples used may differ.

Such extended appreciation is, however, uncommon regarding the overlapping citations discussed above. In multiple cases, the discussion of these citations reappears in a reduced or similar form, if they are explicitly mentioned at all. In four instances, the information presented is approximately identical.[187] This reduced representation is less remarkable when regarding the fact that *Biochemical Individuality* is approximately one hundred pages shorter than *The Human Frontier*, yet presents more sources and examples of individuality. Additionally, *Biochemical Individuality* does not include a secondary sociological discussion of the individualities presented.[188] Per page, therefore, *Biochemical Individuality* presents more than five times as many sources than *The Human Frontier*.[189] This makes briefer descriptions of identical sources inevitable.

Furthermore, *Biochemical Individuality* presents these sources within a different construct of evidence. It presents a mass of examples, many of which are never discussed but merely cited as evidence. They are offered as further reading opportunities for those sceptical of his ideas or as additional scientific literature for those interested in his concept. In contrast, *The Human Frontier* explicitly discusses the examples provided with regard to their social implications and

---

Blood Analyses" and "Normal Young Men", published in 1955 and 1957 respectively, which present his own research (Williams et al., 1955a; Williams, 1957a). These articles do not quote Heath's publication and are therefore not to be viewed as in connection therewith.

187 An analysis of the overlapping citations in *The Human Frontier* and *Biochemical Individuality* can be found in Tab. 1: Comparison Between Overlapping Citations in *The Human Frontier* and *Biochemical Individuality*.

188 *The Human Frontier* cites 108 sources on 301 pages, meaning there is an average of 0.36 citations per page. *Biochemical Individuality*, however, cites 407 sources on 209 pages, averaging 1.95 citations per page.

189 $1.95/0.36 = 5.42$.

potential for change. For the sake of brevity, the full analysis of the overlapping citations between *The Human Frontier* and *Biochemical Individuality* have been summarised in Tab. 1: Comparison Between Overlapping Citations in *The Human Frontier* and *Biochemical Individuality*. *The Human Frontier* presents a plethora of personal descriptions and academic publications presenting evidence of the individuality of every human being. Examples pertinent to the development of BI with special focus on those reappearing in *Biochemical Individuality* have been analysed above. The extensive depiction of each similarity is neither pertinent nor beneficial to this dissertation and has therefore been abstained from.

6.6 Reviews of *The Human Frontier*

An indication of the reception of *The Human Frontier: A New Pathway for Science Toward a Better Understanding of Ourselves* is provided by the published reviews of Williams' work. The following section discusses representative reviews from authors of different faculties in order to portray the academic acceptance of *The Human Frontier* and the theories presented therein. The first discussion-worthy review of *The Human Frontier* is that of Earnest A. Hooton (1887–1954) on the backside of *The Human Frontier*'s cover sleeve. As backside reviews are used to advertise and further recommend a book when a potential costumer holds it in hand, the positive tenor of Hooton's review is to be expected. A well-known physical anthropologist, Hooton describes *The Human Frontier* as "a very important formulation" of the issues covered and as "most stimulating and important" (Hooton, 1946).[190]

Additionally, Williams' book is included in *The Phi Delta Kappan*'s "Selected Bibliography on the Methodology of Educational, Psychological, and Social Research" of 1946/1947 (Good, 1947, p. 148). This collection offers works which "should prove helpful to research workers and graduate students by way of identifying problems for further investigation, in locating critiques of the research in particular fields, and in charting trends in educational, psychological, and social investigation", indicating a positive reception of Williams' work (Good, 1947,

---

190 Ernest A. Hooton was a physical anthropologist, whose regularly published newspaper articles and comments indicate that he was well-known at the time (Garn and Giles, 1995, p. 167). For further information on his life and work, see Garn and Giles (1995).

p. 146). Multiple reviews from various fields, including a minister,[191] a geneticist,[192] and a sociologist,[193] commend Williams for not only presenting examples of individuality, but also for addressing an important issue with regard to the lack of the organised study of individuals (Boisen, 1947, p. 298; Glass, 1947, p. 175; Moore, 1947, p. 238; N., 1947, p. 192). Roger Williams' brother, Robert R. Williams, cites *The Human Frontier* in his article on the most beneficial future focus of scientific research, which largely presents a similar view on the importance of the study of individual humans (Williams, 1948e, p. 119). Nevertheless, these positive appraisals take a minority view in the grand scheme of *The Human Frontier*'s reviews.

The overwhelming majority of critiques have a generally critical tone or are at least critical of multiple aspects of Williams' depictions and theories. Those published by sociologists are the most direct in their criticism and describe only few positive aspects of Williams' work. Though only a small number find fault with his scientific appraisal of individualities in human beings, the sociological aspects of *The Human Frontier* are harshly condemned. Faulted for the "failure to estimate the role of social determinants" (Honigmann, 1947, p. 379),[194] and "insensitivity to culture and its role in social life" (Redfield, 1947, p. 212),[195] "oversimplifications typical of biological determinists, plus a few of his own" (Lee, 1947, p. 211),[196] lead to the accusation of Williams being "entirely innocent of the concept of culture, which is fairly basic" (Moore, 1947, p. 237) and "neglect[ing] social factors to an extent which dismays those trained in the social sciences" (Moore, 1947, p. 237). The overall tone of these critiques suggests that Williams' publication of a book on the facts of social science is seen as an affront. The advice is offered that Williams should have "secure[d] (…) collaboration

---

191   Anton T. Boisen (1876–1965) was an ordained minister and is credited to be "the founder if the clinical pastoral education movement" (Asquith, 1982). For further information on his life and work, see Asquith (1982).
192   Bentley Glass (1906–2005) was a recognised geneticist and promoter of secondary education in the natural sciences. For further information, see Martin (2005).
193   Harry Estill Moore (1897–1966) was a professor of sociology at the University of Texas. For further information, see Smyrl (2023).
194   John J. Honigmann (1914–1977) was an anthropologist at, among others, Yale and Washington State College. For more information, see Honigmann (1982).
195   Robert Redfield (1897–1958) was an anthropologist best known for his work in describing Mexican cultures. For more information, see The Editors of Encyclopaedia Britannica (2022).
196   Alfred McClung Lee (1906–1992) was a noted sociologist at, among others, Yale and New York University. For further information, see Daniels (1992).

in the production of this book", similar to the interdepartmental cooperation he suggests for the development of humanics itself (Moore, 1947, p. 238). This criticism suggests an oversimplification of complex social issues described in *The Human Frontier*, which are doubtlessly seen to stem from Williams' inexperience in the field.

Regarding humanics as the panacea for social issues, similar critique is offered. Described it as being "naive to suppose that a scientific clearing house like humanics will magically realize this goal", *The Human Frontier* is found to overgeneralise the problems of social science in this regard (Honigmann, 1947, p. 379). Such harsh critique is telling, though not surprising. Williams, an outsider in the field of sociology, proposes a very simplistic approach to the solution of problems which the reviewing sociologists and anthropologists have spent their careers studying. This portrayal of a quick fix will, so indicated by the reviews previously cited, have seemed like an attack and insult to those whose work it seemingly aims to make superfluous. This is further indicated by the statement that, "this book can hardly be called a contribution to social science. It is rather a physiologically based psychology of traits applied to problems far beyond its legitimate scope" (Boisen, 1947, p. 298). This aspect is finally demonstrated by the accusation that Williams is seeking "glory" for his idea of humanics and the condemnation of *The Human Frontier* as "an uninformed bit of meddling" (Lee, 1947, p. 211).

The critical analysis of *The Human Frontier* is not limited to its sociological aspects. The extrapolations "made on the basis of extreme deficiencies produced experimentally in animals" is found to have "little bearing on the normal functioning of humans" (Anderson, 1947, p. 444).[197] Additionally, Chapter X, "Humanics and Education", is described as "unsatisfactory and incomplete" (Anderson, 1947, p. 444). The lack of a concrete structure, or suggestion thereof, for humanics is additionally found as wanting (Garside, 1947).[198] Williams' recurring referrals to individualities are, to one reviewer, "spend[ing] an inordinate amount of time on the obvious" (Dempsey, 1946, p. 432).[199] "Insecure choice of

---

197 John E. Anderson (1893–1966) was a psychologist and pioneer in the field of developmental psychology. For more information, see Templin (1968).

198 Edward B. Garside (1907–1999) contributed multiple book reviews in *The New York Times*. A full list of his publications is published online (WorldCat, 2023).

199 Edward W. Dempsey (1911–1975) was an anatomy professor and special assistant for health and medical affairs in Lyndon B. Johnson's Administration in the United States. His life and work are summarised in an obituary by the *New York Times* (The New York Times, 1975).

material and its dubious validity to a hazardous relevancy in its application" is an additional critique-worthy aspect brought forward (The Editors, 1949, p. 158).

As indicated by the reviews presented above, the critique of *The Human Frontier* is manifold. The issue brought up most frequently by critics is that of Williams' inexperience within the fields in which the conclusions and recommendations of *The Human Frontier* are based. Information and proposals are "freely taken outside the author's primary knowledge because – as he naïvely says – the general public (...) is 'more interested in science which accomplishes things than it is in pure science which is devoted to the love of learning for its own sake'" (The Editors, 1949, p. 158). Such free association is seen as ignorance by those reviewing *The Human Frontier* from an expert point-of-view. The overall tenor of these critical reviews is that, though there are some valid aspects of individuality presented, the, in their mind, uninformed extrapolations therefrom invalidate *The Human Frontier* as a whole.

A very different tone is heralded by the multitude of personal letters reaching Williams following *The Human Frontier*'s publication. Collecting quotes and comments from men deemed "important" (Williams, 1947 f), Williams' thesis is accepted most kindly by colleagues of all professions (Camp, 1947; Raible, 1947; Williams, 1947g, 1947h, 1947i). Similarly, a second collection of letters by "famous names" shows a most favourable attitude toward his publication (Baruch, 1946; Clark, 1946; Dulles, 1946; Giannini, 1946; Green, 1947; Holmes, 1946; Hoover, 1946; Huxley, 1957; Jester, 1946; Kettering, 1946; Rockefeller, 1947; Stassen, 1946; Stettinius Jr., 1946). From politicians, military leaders, and doctors to thinkers, religious leaders, journalists, and academics of a variety of fields (one of which adopts *The Human Frontier* as a textbook for his students (Camp, 1947)), Williams' personal correspondence reflects a very different acceptance of his theories. Similarly, reflections on *The Human Frontier* in discussions prior to its release are all favourable toward the general thesis, even if certain specific aspects are criticised. One poet even writes to Williams indicating he has written a poem on his theory of alcoholism (Faller, Undated).

The reception of *The Human Frontier* in the academic community is indicated to be mixed by the published reviews of this work. While the call to increased study of individuals is commended by many, the statements made with regard to the sociological consequences thereof are widely criticised. In all, the tenor of reviews is generally critical. This indicates a lack of general acceptance and a controversy around the topics discussed in *The Human Frontier*.

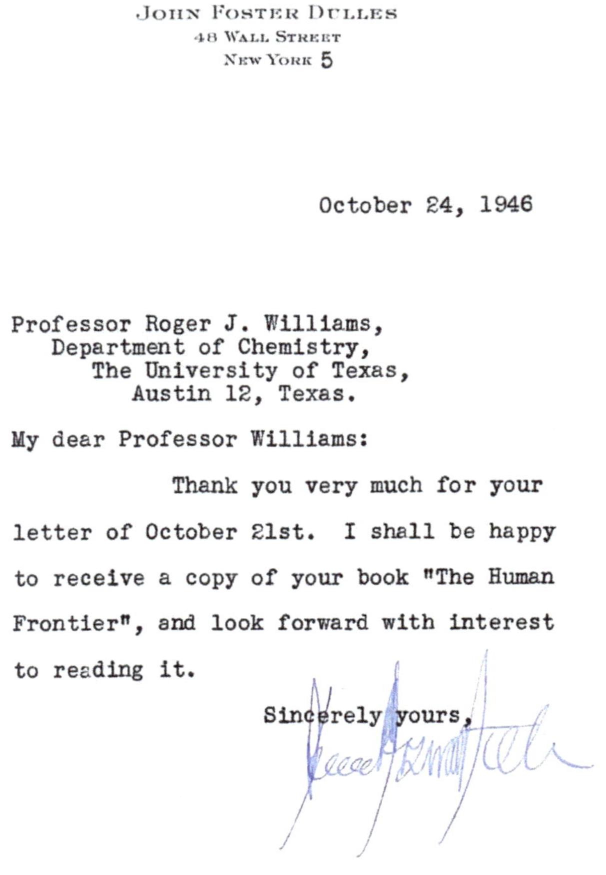

JOHN FOSTER DULLES
48 WALL STREET
NEW YORK 5

October 24, 1946

Professor Roger J. Williams,
Department of Chemistry,
The University of Texas,
Austin 12, Texas.

My dear Professor Williams:

Thank you very much for your letter of October 21st. I shall be happy to receive a copy of your book "The Human Frontier", and look forward with interest to reading it.

Sincerely yours,

Fig. 10: Letter by John Foster Dulles to Williams on 24.10.1946 indicating his interest in *The Human Frontier*; Dolph Briscoe Center for American History, The University of Texas at Austin, Roger Williams Papers, Box 88-087/26a, Folder: Famous Names (Humanics)

APR 28 1947

Room 5600
30 Rockefeller Plaza
New York 20, N.Y.

April 24, 1947.

Dear Mr. Williams:

Thank you for your note of the fifteenth and the copy of your book "The Human Frontier". I appreciate your thought in sending it and look forward to the opportunity of reading it.

It was a pleasure meeting you in Dr. Painter's office the week before last .

Sincerely,

[signature: John D. Rockefeller 3rd]

Mr. Roger J. Williams,
Professor of Chemistry,
The University of Texas,
Austin 12, Texas.

Fig. 11: Letter by John D. Rockefeller III to Williams on 24.04.1947 thanking him for sending a copy of *The Human Frontier* and indicating a meeting two weeks prior; Dolph Briscoe Center for American History, The University of Texas at Austin, Roger Williams Papers, Box 88-087/26a, Folder: Famous Names (Humanics)

6.7 Biochemical Individuality Following *The Human Frontier*

Every individual human being differs from his or her peers in multiple aspects. These individualities, which are often based upon hereditary metabolic differences, are at the root of most of the social issues of our time. These individualities must be studied further, so that society can improve on its possibilities of social control and provide ideally for each individual in all aspects of his or her life. Medical and psychological research on metabolic individualities could help to understand a multitude of diseases and improve the care provided to patients. The research of individuals must be at the forefront of a new science to be called humanics.

6.8 Conclusion

As a call for increased research on human individuals, *The Human Frontier: A New Pathway for Science Toward a Better Understanding of Ourselves is* aimed at the general public. Social control, which *The Human Frontier* seeks to improve, refers to the improved social integration of individuals according to their unique needs and predispositions. The examples of individuality presented range from personal accounts and uncited descriptions to cited scientific evidence. Williams' discussion of theories posed previously by other scientists offers insight into external influences on his research. Multiple overlapping citations between *The Human Frontier* and *Biochemical Individuality: The Basis for the Genetotrophic Concept* indicate that the former can be considered a precursor of the latter. The published reviews of *The Human Frontier* indicate a mixed and largely lacking acceptance within the scientific community, especially from anthropological and sociological faculties. *The Human Frontier* is Roger Williams' first publication which sets out to categorically prove and provide explicit examples for the individuality of every human being.

7 The Scientific Study of Individuals and Alcoholism

The publication of *The Human Frontier: A New Pathway for Science Toward a Better Understanding of Ourselves* (*The Human Frontier*) represents a turning point with regard to the development of BI. The first scientific review conceptualised to explore individualities between human beings is published by Roger Williams the following year. While, as a sociological book, *The Human Frontier* is outside of Williams' field of expertise, the biochemical research on individuality is entirely within the realm of his vocational training. Generally speaking, the nature of the research published by Williams is reformed following his public call for increased study of individualities. In keeping with the ideas presented in *The Human Frontier*, Williams almost exclusively publishes academic papers exploring different aspects of individuality. This chapter will discuss all papers and articles relevant to the development of BI published by Williams from 1946 up to and including December of 1949. The articles discussed have been selected according to their content, placing special focus on those revealing new evidence of individuality researched by Williams himself. Papers presenting similar or identical pieces of evidence are mentioned, yet no further analysis is provided as they are reiterations of content discussed elsewhere.

7.1 "Biochemical Individuality and Its Implications"[200]

The first relevant article discussing themes of individuality that appears after the publication of *The Human Frontier*, entitled "Humanics: A Crucial Need", acts as a preview to the topics covered in the aforementioned book and can therefore be categorised as a promotional publication (Williams, 1946k, 1947b). All but one source cited in this article are also quoted in *The Human Frontier*, and the one source which is not referenced in Williams' (1947b, Citation 3) previous publications later reappears in *Biochemical Individuality: The Basis for the Genetotrophic Concept* (Williams, 1956a, Chap. X Citation 33). This citation is the solitary novel aspect "Humanics: A Crucial Need" offers. Similarly, "Will Science Meet a New Challenge?" presents no new information (Williams, 1947d), though it does

---

200 (Williams, 1947j) To improve the clarity of this section-title and avoid confusion, the citation for this quote is presented in this footnote, deviating from the citation style used in the rest of this dissertation.

supply evidence of the context in which Williams' strides towards humanics take place. He repeatedly speaks of the possibilities of "super-destructive nuclear war" (Williams, 1947d, p. 282), concluding that "atomic disintegration becomes dangerous only when it is under [human] control" (Williams, 1947d, p. 283). It is published in *American Scientist*, a well-known popular science Magazine,[201] two years after the bombings of Hiroshima and Nagasaki at the end of the Second World War.[202] The subsequent horror and fear of nuclear warfare, which accompanied all nations of the world throughout the Cold War, features as a prominent selling point of Williams' plans for the interdepartmental study of human beings.[203] According to Williams, "ultimately preventing international war" constitutes one of the realistic potentials of humanic research (Williams, 1947d, p. 286). The prominent nature of *American Scientist* additionally supports the claim that Williams aimed to address as widespread an audience as possible in his pursuit of intensified individualised research.[204]

"Biochemical Individuality and Its Implications" (Williams, 1947j), an article also appearing in *Chemical and Engineering News* in April of 1947, is similar to those discussed previously in the majority of topics it covers. However, two aspects of this piece increase its relevance to BI and its development. Primarily, this article contains the first appearance of the conjugate term "Biochemical Individuality". While terms such as "individual variations" "distinctive metabolic traits", or "individual metabolic idiosyncrasies" previously appearing in publications can be considered synonymous, the idiom "Biochemical Individuality" is emblematic to the theories encased in the identically named publication of 1956.[205] The semantics of this first conjugation of the terms "biochemical" and "individuality" are of additional interest. In the preface to *Biochemical Individuality* Williams speaks of the nature of his first thoughts on individuality:

> When my interest in this area [Biochemical Individuality] first developed, I regarded it as considerably divergent from my chosen field of research interest – biochemistry. However, as time has gone on and research results have accumulated, it has become clearer to me that individuality and applied biochemistry are inextricably intertwined. I no longer regard my interest in individuality as a departure from biochemistry. (Williams, 1956a, p. x)

---

201   See Footnote 139.
202   For further information of the History of WWII, see Keegan (2005).
203   For further information on the general history of the Cold War, see Westad (2017).
204   See Section 6.2. Audience.
205   See Sections 4.5. "'Taste Deficiency' for Creatine", 5.2. The Vitamin Content of Tissues, and 6.4. "Distinctive Metabolic Traits".

The growing acceptance described in 1956 is here signified by the establishment of the term "Biochemical Individuality". It is additionally noteworthy that the term "humanics" does not appear in this article. The accurate study of human beings still remains a primary concern to Williams and his colleagues and is often described at great lengths. "Will Science Meet a New Challenge", however, is the last publication in which Roger Williams uses this term to describe such research until it reappears in a contribution to a symposium published in 1954 (Williams, 1954e, p. 328). Though the interdepartmental study of individuals continues to be promoted, an indication for the discontinuation of the term "humanics" is offered only in a later book, *Free and Unequal*. Here, Williams describes how "a prominent man" rejects his concept of individuality for the sole reason of disliking the term "humanics" (Williams, 1953a, p. 154). Following this comment, the term ceases to appear for several years.

Simultaneously to this change, the biochemical aspects of individuality become more of a focus of Williams' articles and research. Differences in the rudimentary metabolic makeup of every organism, which are based on genetic variations, are supposed as the source of all individuality. The integration of the study of individuals into biochemistry no longer necessitates the use of the term "humanics". While the physiological effects of such individualities remain to be of interest, the detailed biochemical analysis of these individualities must come first. Williams' previous publications regarding individuality, especially *The Human Frontier*, mainly focus on aspects outside of his field of expertise. From this point onwards, the research of individuality is primarily focused on its biochemical aspects.

The term BI appears in most subsequent articles. Furthermore, "Biochemical Individuality and Its Implications" calls increased attention to the individualities of human behaviour and illness with relation to alcoholism. In addition to citing an article later reused in *Biochemical Individuality*,[206] Williams calls attention to his own first concrete study of individualities of humans discussed elsewhere (Williams, 1947j, p. 1113).[207] "Biochemical Individuality and Its Implications" additionally functions as the first of a series titled "Biochemical Individuality". A total of five publications from DBUT appear therein, discussing different aspects of BI. The consequent four articles, not all featuring Williams as an author, appear in *Archives of Biochemistry* using a uniform numbering system.

---

206 See Tab. 2: Articles Cited by *Biochemical Individuality* Appearing in Prior Publications.
207 See Section 7.3. "Alcoholics and Metabolism".

This first article is divergent in this aspect, though "Biochemical Individuality II" explicitly names "Biochemical Individuality and its Implications" as the first of this series of studies (Thompson and Kirby, 1949, p. 210). It therefore signifies the kick-off to DBUT's research into Biochemical Individuality, after which the research of individuals begins to gain momentum.

Tab. 2: Articles Cited by *Biochemical Individuality* Appearing in Prior Publications

No.	Citation in *Biochemical Individuality*	Page(s) of Citation in *Biochemical Individuality*	Other Publications Containing the Identical Citation	Page(s) of Citation in Other Publications
1	Clark W. Heath et al., What People Are, Harvard University Press, Cambridge, Mass., 1945.	28, 35, 52, 120	The Human Frontier (1946)	11
2	Leo Loeb, The Biological Basis of Individuality, Charles C. Thomas, Springfield. Ill., 1947.	132	The Human Frontier (1946)	26
			The Genetotrophic Concept – Nutritional Deficiencies and Alcoholism (1954)	795
3	Ralph S. Banay, Quart]. Studies Ale., 4, 580–605, 1944.	108	The Human Frontier (1946)	33
			Biochemical Genetics and its Human Implications (1956)	175
4	R. W. Engel, Proc. Soc. Exptl. Biol. Med., 52, 281–282, 1943.	156	The Human Frontier (1946)	38
			Biochemical Individuality III. Genetotrophic Factors in the Etiology of Alcoholism (1949)	280
			Genetotrophic Diseases; Alcoholism (1950)	243
			Human Nutrition and Individual Variability (1956)	19

Tab. 2: Continued

No.	Citation in *Biochemical Individuality*	Page(s) of Citation in *Biochemical Individuality*	Other Publications Containing the Identical Citation	Page(s) of Citation in Other Publications
5	R. F. Light and L. J. Cracas, Science, 87, 90, 1938.	149	The Human Frontier (1946)	38
			Biochemical Individuality III. Genetotrophic Factors in the Etiology of Alcoholism (1949)	280
			Genetotrophic Diseases; Alcoholism (1950)	243
			Human Nutrition and Individual Variability (1956)	17
6	C. J. Warden, H. C. Brown, and Sherman Rose, J. Exptl. Psychology, 35, 57–70, 1945.	125	The Human Frontier (1946)	46
7	Ward C. Halstead, Science, 101, 615–616, 1945.	124	The Human Frontier (1946)	46
8	Albert F. Blakeslee, Science, 81, 504–507, 1935.	127	The Human Frontier (1946)	71
9	Albert F. Blakeslee and Theodora Nussman Salmon, Proc. Natl. Acad. Sci. US., 21, 84–90, 1935.	127	The Human Frontier (1946)	71
			Implications of Humanics for Law and Science (1954)	334
10	A. F. Blakeslee and A. L. Fox, J. Heredity, 23, 97–106, 1932.	130	The Human Frontier (1946)	74
11	Edgar A. Hines, Jr., and George E. Brown, Am. Heart J., 11, 1–9, 1936.	125	The Human Frontier (1946)	83

(continued)

Tab. 2: Continued

No.	Citation in *Biochemical Individuality*	Page(s) of Citation in *Biochemical Individuality*	Other Publications Containing the Identical Citation	Page(s) of Citation in Other Publications
12	N. Kleitman, Sleep and Wakefulness, University of Chicago Press, Chicago, Ill., 1939.	122	The Human Frontier (1946)	92
13	Arthur Grollman, Essentials of Endocrinology, J. B. Lippincott Co., Philadelphia, Pa., 2nd ed., 1947.	81	The Human Frontier (1946)	110
			An Introduction to Biochemistry (1948)	399
			The Genetotrophic Approach to Alcoholism (1954)	201
14	Max A. Goldzieher, The Endocrine Glands, D. Appleton-Century Co., New York, N. Y. and London, England, 1939.	90	The Human Frontier (1946)	113
			The Genetotrophic Approach to Alcoholism (1954)	201
15	A. T. Rasmussen, Am. J. Anal., 42, 1–27, 1928.	91, 92	The Human Frontier (1946)	116
			The Genetotrophic Approach to Alcoholism (1954)	201
16	A. T. Rasmussen, Endocrinology, 8, 509–524, 1924.	91, 92	The Human Frontier (1946)	116
			The Genetotrophic Approach to Alcoholism (1954)	201
17	A. T. Rasmussen, Endocrinology, 12, 129–150 (1928).	91, 92	The Human Frontier (1946)	116
			The Genetotrophic Approach to Alcoholism (1954)	201

Tab. 2: Continued

No.	Citation in *Biochemical Individuality*	Page(s) of Citation in *Biochemical Individuality*	Other Publications Containing the Identical Citation	Page(s) of Citation in Other Publications
18	Oscar Riddle, Endocrines and Constitution in Doves and Pigeons, Carnegie Institution of Washington, Washington, D.C., 1947.	19, 92	The Human Frontier (1946)	120
19	Advertisement in J. Am. Med. Assoc., Oct., 20, 1945.	143	The Human Frontier (1946)	230
20	Louis Goodman and Alfred Gilman, The Pharmacological Basis of Therapeutics, The Macmillan Co., New York, N.Y., 1941.	111	The Human Frontier (1946)	232
			An Introduction to Biochemistry (1948)	466
21	George Draper, C. W. Dupertuis, J. L. Caughey, Human Constitution in Clinical Medicine, Paul B. Hoeber, Inc., New York, N.Y., 1944.	2, 190	The Human Frontier (1946)	233, 243
22	Samuel Henry Kraines, The Therapy of the Neuroses and Psychoses, Lea and Febiger, Philadelphia, Pa., 2nd ed., 1945.	201	The Human Frontier (1946)	235

(continued)

Tab. 2: Continued

No.	Citation in *Biochemical Individuality*	Page(s) of Citation in *Biochemical Individuality*	Other Publications Containing the Identical Citation	Page(s) of Citation in Other Publications
23	Edgar S. Gordon, "Pantothenic Acid in Human Nutrition," in E. A. Evans, Jr., ed., The Biological Action of the Vitamins, The University of Chicago Press, Chicago, 111., 1942	154	The Human Frontier (1946)	247
			An Introduction to Biochemistry (1948)	286
24	W. W. Jetter, Am. J. Med. Sci., 196, 475 (1938).	108	Biochemical Individuality and Its Implications (1947)	1112
			The Etiology of Alcoholism: A Working Hypothesis Involving the Interplay of Hereditary and Environmental Factors (1947)	576
			Implications of Humanics for Law and Science (1954)	342
25	H.C. Sherman and H. L. Campbell, Proc Natl. Acad. Sci. U. S., 31, 164–166, 1945.	144	Humanics: A Crucial Need (1947)	178
26	Erwin E. Nelson, J. Am. Med. Assoc., 113, 1373–1375, 1939.	114	The Etiology of Alcoholism: A Working Hypothesis Involving the Interplay of Hereditary and Environmental Factors (1947)	571
27	P. T. Young, Psychology Bull., 38, 129–164, 1941.	129	The Etiology of Alcoholism: A Working Hypothesis Involving the Interplay of Hereditary and Environmental Factors (1947)	572

Tab. 2: Continued

No.	Citation in *Biochemical Individuality*	Page(s) of Citation in *Biochemical Individuality*	Other Publications Containing the Identical Citation	Page(s) of Citation in Other Publications
28	Curt P. Richter, Quart. J. Studies Alc., 1, 650–662, 1941.	109, 218	The Etiology of Alcoholism: A Working Hypothesis Involving the Interplay of Hereditary and Environmental Factors (1947)	574
			Implications of Humanics for Law and Science (1954)	334
29	J. Warkentin, L. Warkentin, and A. C. Ivy, Am. J. Psychol., 139, 139–146, 1943.	129	The Etiology of Alcoholism: A Working Hypothesis Involving the Interplay of Hereditary and Environmental Factors (1947)	574
30	John M. Nagle, J. Allergy, 10, 179–181, 1939.	108	The Etiology of Alcoholism: A Working Hypothesis Involving the Interplay of Hereditary and Environmental Factors (1947)	577
			Biochemical Genetics and its Human Implications (1956)	172
31	Douglas McG. Kelley and S. Eugene Barrera, Psychiat. Quart., 15, 224–248, 1941.	109	The Etiology of Alcoholism: A Working Hypothesis Involving the Interplay of Hereditary and Environmental Factors (1947)	577
32	E. M. P. Widmark, Physiological Papers Dedicated to Professor August Krogh, Levin & Munksgaard, Copenhagen, Denmark, 1926.	109	The Etiology of Alcoholism: A Working Hypothesis Involving the Interplay of Hereditary and Environmental Factors (1947)	577
			The Genetotrophic Approach to Alcoholism (1954)	206

(*continued*)

Tab. 2: Continued

No.	Citation in *Biochemical Individuality*	Page(s) of Citation in *Biochemical Individuality*	Other Publications Containing the Identical Citation	Page(s) of Citation in Other Publications
33	Robert Fleming. "Medical Treatment of the Inebriate," in Alcohol, Science and Society, Journal of Studies on Alcohol, Inc., New Haven, Conn., 1945, p. 391.	109, 173	The Etiology of Alcoholism: A Working Hypothesis Involving the Interplay of Hereditary and Environmental Factors (1947)	579
			Genetotrophic Diseases; Alcoholism (1950)	253
34	J. Mardones, N. Segovia, and E. Onfray, Arch. Biochem., 9, 401–406, 1946.	160	Individual Metabolic Patterns, Alcoholism, Genetotrophic Diseases (1949)	265
			Biochemical Individuality III. Genetotrophic Factors in the Etiology of Alcoholism (1949)	278
			Genetotrophic Diseases; Alcoholism (1950)	241
			The Genetotrophic Concept – Nutritional Deficiencies and Alcoholism (1954)	803
			The Genetotrophic Approach to Alcoholism (1954)	203, 204
			Dietary Deficiencies in Animals in Relation to Voluntary Alcohol and Sugar Consumption (1955)	234
			Voluntary Alcohol Consumption by Rats Following Administration of Glutamine (1955)	503

Tab. 2: Continued

No.	Citation in *Biochemical Individuality*	Page(s) of Citation in *Biochemical Individuality*	Other Publications Containing the Identical Citation	Page(s) of Citation in Other Publications
35	Herschel B. Mitchell and Mary B. Houlahan, Am. J. Botany, 31–35, 1946.	10	Biochemical Individuality III. Genetotrophic Factors in the Etiology of Alcoholism (1949)	275
			Genetotrophic Diseases; Alcoholism (1950)	239, 243
			The Concept of Genetotrophic Disease (1950)	287
			The Genetotrophic Concept – Nutritional Deficiencies and Alcoholism (1954)	794
			The Genetotrophic Approach to Alcoholism (1954)	196
36	Roscoe A. Brady and W. W. Westerfeld, Quart. J. Studies Alc., 7, 499–505, 1947.	160	Biochemical Individuality III. Genetotrophic Factors in the Etiology of Alcoholism (1949)	278
			Genetotrophic Diseases; Alcoholism (1950)	241
			The Genetotrophic Concept – Nutritional Deficiencies and Alcoholism (1954)	803
			The Genetotrophic Approach to Alcoholism (1954)	203, 204
			Voluntary Alcohol Consumption by Rats Following Administration of Glutamine (1955)	503

(continued)

Tab. 2: Continued

No.	Citation in *Biochemical Individuality*	Page(s) of Citation in *Biochemical Individuality*	Other Publications Containing the Identical Citation	Page(s) of Citation in Other Publications
37	Daniel Melnick, William D. Robinson, and Henry Field, Jr., J. Biol. Chem., 136, 131–144, 1940.	103	The Biochemistry of B Vitamins (1950)	55
38	G. W. Beadle and E. L. Tatum, Proc. Natl. Acad. Sci. U.S., 27, 499–506, 1941.	10	The Biochemistry of B Vitamins (1950)	86
39	Roy C. Thompson and Helen M. Kirby, Arch. Biochem., 21, 210–216, 1949.	99	The Concept of Genetotrophic Disease (The Lancet, 1950)	287
			Concept of Genetotrophic Disease (Nutrition Reviews, 1950)	257
			Biochemical Individuality. V. Explorations with Respect to the Metabolic Patterns of Compulsive Drinkers (1950)	28
			"Introduction, General Discussion and Tentative Conclusions" in Biochemical Institute Studies IV: Individual Metabolic Patterns and Human Disease: An Exploratory Study Utilizing Predominantly Paper Chromatographic Methods	13
			The Genetotrophic Concept – Nutritional Deficiencies and Alcoholism (1954)	803

Tab. 2: Continued

No.	Citation in *Biochemical Individuality*	Page(s) of Citation in *Biochemical Individuality*	Other Publications Containing the Identical Citation	Page(s) of Citation in Other Publications
40	F. G. Harbaugh and Joe Dennis, Am. J. Vet. Research, 8, 396–399, 1947.	138	Concept of Genetotrophic Disease (Nutrition Reviews, 1950)	259
41	Borden's Review of Nutritional Research, 8, 6, 1947.	136	Biochemical Individuality. V. Explorations with Respect to the Metabolic Patterns of Compulsive Drinkers (1950)	29
			"Introduction, General Discussion and Tentative Conclusions" in Biochemical Institute Studies IV: Individual Metabolic Patterns and Human Disease: An Exploratory Study Utilizing Predominantly Paper Chromatographic Methods	14
42	Barry J. Anson, Atlas of Human Anatomy, W.V. Saunders Co., Philadelphia, Pa. and London, England, 1950.	22 and throughout Chapter 3	Implications of Humanics for Law and Science (1954)	332
			Biochemical Approach to the Study of Personality (1954)	31
43	Marion M. Maresh, Pediatrics, 2, 382–404, 1948.	28	Implications of Humanics for Law and Science (1954)	332
44	K. S. Lashley, Psychological Reviews, 54, 333–334, 1947.	44	Implications of Humanics for Law and Science (1954)	333
			Biochemical Approach to the Study of Personality (1954)	31

(*continued*)

Tab. 2: Continued

No.	Citation in *Biochemical Individuality*	Page(s) of Citation in *Biochemical Individuality*	Other Publications Containing the Identical Citation	Page(s) of Citation in Other Publications
45	Harry Harris, An Introduction to Human Biochemical Genetics, Cambridge University Press, London, England and New York, N.Y., 1953.	99	Implications of Humanics for Law and Science (1954)	335
46	A. F. Blakeslee, Science, 48, 298–299, 1918.	130	Implications of Humanics for Law and Science (1954)	335
47	Seymour S. Kety, Biology of Mental Health and Disease, Paul B. Hoeber, Inc., New York, N. Y., 1952.	199, 201	Implications of Humanics for Law and Science (1954)	336
48	John W. Gowen, Am. J. Human Gen., 4, 285–302, 1952.	170	The Genetotrophic Concept – Nutritional Deficiencies and Alcoholism (1954)	795
49	Irvine H. Page, Esben Kirk, William H. Lewis, Jr., William R. Thompson, and Donald D. Van Slyke, J. Biol. Chem., Ill, 638, 1935.	53, 55, 57, 58	The Genetotrophic Concept – Nutritional Deficiencies and Alcoholism (1954)	797
			The Genetotrophic Approach to Alcoholism (1954)	197
50	Harry Eldon Sutton, Univ. Texas Publ., 5109, 173–180, 1951.	99	The Genetotrophic Concept – Nutritional Deficiencies and Alcoholism (1954)	803

Tab. 2: Continued

No.	Citation in *Biochemical Individuality*	Page(s) of Citation in *Biochemical Individuality*	Other Publications Containing the Identical Citation	Page(s) of Citation in Other Publications
51	Helen Kirby Berry and Louise Cain, Univ. Texas Publ., 5109, 165–172, 1951.	99	The Genetotrophic Concept – Nutritional Deficiencies and Alcoholism (1954)	803
			Urinary Amino Acids, Creatinine and Phosphate in Muscular Dystrophy (1955)	385
52	Janet G. Reed, Univ. Texas Publ., 5109, 144–150, 1951.	101, 173, 191	The Genetotrophic Concept – Nutritional Deficiencies and Alcoholism (1954)	803
			The Genetotrophic Approach to Alcoholism (1954)	203
			Dietary Deficiencies in Animals in Relation to Voluntary Alcohol and Sugar Consumption (1955)	234
53	Janet G. Reed, Univ. Texas Publ., 5109, 139–143, 1951.	100, 143, 173	The Genetotrophic Concept – Nutritional Deficiencies and Alcoholism (1954)	803
			Dietary Deficiencies in Animals in Relation to Voluntary Alcohol and Sugar Consumption (1955)	234
			Voluntary Alcohol Consumption by Rats Following Administration of Glutamine (1955)	505

(continued)

Tab. 2: Continued

No.	Citation in *Biochemical Individuality*	Page(s) of Citation in *Biochemical Individuality*	Other Publications Containing the Identical Citation	Page(s) of Citation in Other Publications
54	Ernest Beerstecher, Jr., Janet G. Reed, William Duane Brown, and L. Joe Berry, Univ. Texas Publ. 5109, 115–138, 1951.	173	The Genetotrophic Concept – Nutritional Deficiencies and Alcoholism (1954)	804
			The Genetotrophic Approach to Alcoholism (1954)	204
			Dietary Deficiencies in Animals in Relation to Voluntary Alcohol and Sugar Consumption (1955)	234
55	Icie G. Macy, Nutrition and Chemical Growth in Childhood, Charles C. Thomas, Springfield, 111., and Baltimore, Md., Vol. I, 1942.	103, 137	The Genetotrophic Approach to Alcoholism (1954)	202
			Dietary Deficiencies in Animals in Relation to Voluntary Alcohol and Sugar Consumption (1955)	241
56	MJ. Mardones, Quart. J. Studies Ale., 12, 563–575, 1951.	162	The Genetotrophic Approach to Alcoholism (1954)	203, 204
			Dietary Deficiencies in Animals in Relation to Voluntary Alcohol and Sugar Consumption (1955)	234
57	E. Rissel and F. Wewalka, Klin. Wochschr., 30, 1065–1069, 1952.	61	The Genetotrophic Approach to Alcoholism (1954)	197
58	E. Rissel and F. Wewalka, Klin. Wochschr., 30, 1069–1073, 1952.	61	The Genetotrophic Approach to Alcoholism (1954)	197
59	Leland C. Clark, Jr., and Elizabeth Beck, J. Pedia., 36, 335–341, 1950.	69	The Genetotrophic Approach to Alcoholism (1954)	200

Tab. 2: Continued

No.	Citation in *Biochemical Individuality*	Page(s) of Citation in *Biochemical Individuality*	Other Publications Containing the Identical Citation	Page(s) of Citation in Other Publications
60	Alton Meister, J. Clin. Invest., 27, 263–271, 1948.	70	The Genetotrophic Approach to Alcoholism (1954)	200
61	Jules Tuba, Max M. Cantor, and Herman Siemens, J. Lab. Clin. Med., 32, 194–195, 1947.	70	The Genetotrophic Approach to Alcoholism (1954)	200
62	Leland C. Clark, Jr., and Elizabeth I. Beck, J. Applied Physiol., 2, 343–347, 1949.	71	The Genetotrophic Approach to Alcoholism (1954)	200
63	Arthur Sawitsky, Howard M. Fitch, and Leo M. Meyer, J. Lab. Clin. Med., 33, 203–206, 1948.	71	The Genetotrophic Approach to Alcoholism (1954)	200
64	G. E. Hall and C. C. Lucas, J. Pharmacol, and Exptl. Therap., 61, 10–20, 1937.	71	The Genetotrophic Approach to Alcoholism (1954)	200
65	Michael Somogyi, Arch. Internal Med., 67, 665–679, 1941.	72	The Genetotrophic Approach to Alcoholism (1954)	200
66	Roger S. Dille and Charles H. Watkins, J. Lab. Clin. Med., 33, 480–486, 1948.	72	The Genetotrophic Approach to Alcoholism (1954)	200
67	Charles Huggins and Dwight Raymond Smith, J. Biol. Chem., 170, 391–398, 1947.	73	The Genetotrophic Approach to Alcoholism (1954)	200

(continued)

Tab. 2: Continued

No.	Citation in *Biochemical Individuality*	Page(s) of Citation in *Biochemical Individuality*	Other Publications Containing the Identical Citation	Page(s) of Citation in Other Publications
68	Mandred W. Comfort and Arnold E. Osterberg, Med Clinics N. Amer., 24, 1137–1149, 1940.	73	The Genetotrophic Approach to Alcoholism (1954)	200
69	Elijah Adams, Mary McFadden, and Emil L. Smith, J. Biol. Chem., 198, 663–670, 1952.	73	The Genetotrophic Approach to Alcoholism (1954)	200
70	W. H. Fishman, M. Smith, D. B. Thompson, C. D. Bonner, S. C. Kasdon, and F. Homburgcr, J. Clin. Invest., 30, 685, 1951.	74	The Genetotrophic Approach to Alcoholism (1954)	200
71	Harold A. Harper, Maxine E. Hutchin, and Joe R. Kimmel, Proc. Soc. Exptl. Biol. Med., 80, 768–771, 1952.	55	The Genetotrophic Approach to Alcoholism (1954)	197
72	Errett C. Albritton, ed., Standard Values in Blood, W. B. Saunders Co., Philadelphia, Pa., and London, England, 1952.	2, 4, 7, 33, 50, 52, 135	The Genetotrophic Approach to Alcoholism (1954)	197
73	Gregory Pincus and Kenneth V. Thimann, eds. The Hormones, Academic Press, Inc., New York, N. Y., 1948, Vol. I.	85	The Genetotrophic Approach to Alcoholism (1954)	201

Tab. 2:   Continued

No.	Citation in *Biochemical Individuality*	Page(s) of Citation in *Biochemical Individuality*	Other Publications Containing the Identical Citation	Page(s) of Citation in Other Publications
74	H. S. Anker, J. Biol. Chem., 176, 1337–1352, 1948.	77	Dietary Deficiencies in Animals in Relation to Voluntary Alcohol and Sugar Consumption (1955)	241
75	Martha F. Trulson, Robert Fleming, and Frederick J. Stare, J. Am. Med. Assoc., 155, 114–119, 1954.	185, 205	Dietary Deficiencies in Animals in Relation to Voluntary Alcohol and Sugar Consumption (1955)	242

7.2 "The Etiology of Alcoholism"[208]

"The Etiology of Alcoholism: A Working Hypothesis Involving the Interplay of Hereditary and Environmental Factors" is an abridged summary of research regarding the possibility that alcoholism may be of multifactorial origin (Williams, 1947e). Presenting a theory much akin to the "Diathesis-Stress-Model" still applied to psychiatric illnesses today,[209] Williams offers a review of alcoholism research, much like *The Human Frontier* reviews studies implying the individuality of humans.[210] The first pages of "The Etiology of Alcoholism: A Working Hypothesis Involving the Interplay of Hereditary and Environmental Factors" present general examples of Biochemical Individuality, most of which

---

[208] (Williams, 1947e). To improve the clarity of this section-title and avoid confusion, the citation for this quote is presented in this footnote, deviating from the citation style used in the rest of this dissertation.

[209] Following definition of this model is widely accepted: "The diathesis-stress model describes how genetic or biological factors interact with environmental stress which results in a disorder or condition" (Goforth et al., 2011, p. 502). This model for the aetiology of multiple psychological diseases, such as schizophrenia and anxiety disorders, is still taught in the Psychiatric Institute of the Rheinische Friedrich-Wilhelms-Universität Bonn. In some cases, the term vulnerability-stress-model is used synonymously. For further information, see Goforth et al. (2011), Kendler (2020), and Salomon and Jin (2013).

[210] See Chapter 6. *The Human Frontier*.

already appear in *The Human Frontier*.[211] These provide the basis for the hypothesis this paper presents. The article is published in the *Quarterly Journal of Studies on Alcohol* and is therefore technical in its approach and phrasing.

The central hypothesis of "The Etiology of Alcoholism" considers the disease as multifactorial, with genetic and environmental factors being equally relevant in its aetiology. It thereby suggests that alcoholism is more likely to occur when a "metabolic individuality (…) predisposes toward addiction", though this genetically determined individuality alone does not suffice to make an individual an alcoholic (Williams, 1947e, p. 576). Among others, the environment of the individual as well as the availability and social acceptance of alcohol also play a crucial role in creating an alcoholic (Williams, 1947e, p. 581). The largest section of this article provides examples of individuality regarding humans' reaction to alcohol. Here, much like *The Human Frontier*, multiple sources of "The Etiology of Alcoholism" are recited in *Biochemical Individuality*. The full list of recurring sources, including their respective pages of citation, can be found in Tab. 2: Articles Cited by *Biochemical Individuality* Appearing in Prior Publications.

One of these sources is of such relevance to Williams that it warrants two separate citations in *Biochemical Individuality*. The article in question, by Robert Fleming, describes the case of an apparent alcoholic, drinking large amounts of whisky daily yet successfully managing an important business and living to the age of 93 (Fleming, 1945).[212] Used as an example for individuality with regard to chronic consumption of alcohol on page 109 and as evidence of individuality in nutrition on page 173, Williams uses virtually identical formulations to describe this example in both citations in *Biochemical Individuality* (Williams, 1956a). In "The Etiology of Alcoholism", the same example is used to illustrate

---

211 The examples of Garrod's metabolic idiosyncrasies (p. 569), Beadle's genetic research (p. 569), Loeb's research on tissue transplantation (p. 571) (all three, see Section 6.4. "Distinctive Metabolic Traits"), odour detection by bloodhounds (p. 571) (see Section 5.3. Practical Applications of Vitamin Research and Section 6.5. "Individuals Vary Greatly"), drug idiosyncrasies (p. 571) (see Sections 4.3. Adverse Drug Reaction, 4.5. "'Taste Deficiency' for Creatine", 5.3. Practical Applications of Vitamin Research, and 6.5. "Individuals Vary Greatly"), and taste deficiency for PTC (p. 575) (see Section 6.5. "Individuals Vary Greatly") all reappear (Williams, 1947e).

212 Dr. Robert E. Fleming was a psychiatrist at Harvard University in Boston and expert on alcoholism. He published multiple studies on the treatment and origin of alcoholic craving in well-respected journals (Fleming, 1945). At the time of writing, no further biographical information is available.

individualities with regard to alcohol craving (Williams, 1947e, p. 579). The difference in contexts, in which identical examples are presented, demonstrate the change of focus with regard to Williams' publications on individuality over time. While such instances of uniqueness are still sought to provide justification for the research of alcoholism in 1947, they are drawn upon to confirm his theories in 1956. This example is additionally noticeable due to the great similarity in the language of these three separate accounts.

The most striking similarity between "The Etiology of Alcoholism" and *Biochemical Individuality* involves multiple sources and more extensive overlap. Pages 108 and 109 of *Biochemical Individuality* and pages 576 to 578 in "The Etiology of Alcoholism" are nearly identical in their presentation, both in their explicit content and wording. Both present evidence of the individuality of alcohol metabolism in humans, addressing the identical sources and the same uncited examples of pylorospasm and vomiting following alcohol intake as well as the varying symptoms of acute alcohol intoxication between individuals (Williams, 1947e, pp. 577–578, 1956a, p. 109).[213] This indicates that "Etiology of Alcoholism" and the research therein serve as a knowledge base for the section on alcohol in *Biochemical Individuality*. Additionally, this overlap indicates a continued relevance of alcoholism and its treatment to BI, supporting the hypothesis of a continual and steady build-up of evidence leading to the publication of *Biochemical Individuality* in 1956.

The lack of new information in *Biochemical Individuality* could be considered as indicating a lack of progress regarding the study of alcoholism in the timespan between the two publications or a lack of repeated in-depth research on the topic by Williams. The stark similarities in the structure of the two excerpts could additionally be understood to indicate a simple replication of the knowledge from the former to the latter without further exploration. Merely one source not already present in "The Etiology of Alcoholism" is cited within the section on alcohol in *Biochemical Individuality*, further supporting this claim.[214] However, the discussion of alcohol is embedded in the chapter "Pharmacological Manifestations" in *Biochemical Individuality* and covers two pages within the seven-page chapter (Williams, 1956a, chap. VIII). While the lack of an own section discussing

---

213   Though in slightly altered order, *Biochemical Individuality* and "The Etiology of Alcoholism" both discuss the identical sources by Jetter, Nagle, Kelley, Widmark, Richter, Kraines, and Banay (Williams, 1947e, pp. 576–578, 1956b, pp. 108–109). Merely the citation of Fleming's work in *Biochemical Individuality* differentiates the two sections (Williams, 1956a, p. 109).

214   It is cited in Chapter VIII as Source 15: Charles C. Hewitt, Quart. J. Studies Alc., 4, 368–386 (1943) (Williams, 1947e, References, 1956a, pp. 108–110, 118).

its relevance could indicate a lack of progress regarding the knowledge of individualities with regard to alcohol, Williams' extensive publication on the topic of alcoholism may have played a role in its lack of far-reaching treatment in *Biochemical Individuality*. With the research of alcoholism at the forefront of Williams' investigations in the years following "The Etiology of Alcoholism", including the publication of a book on the subject, a lack of knowledge regarding the developments in the field of alcoholism is unlikely.[215] Therefore, it can be concluded that the summary provided by "The Etiology of Alcoholism" was simply deemed adequate by Williams to conclusively present individualities regarding the various reactions to alcohol, even nine years following its original publication.

An important aspect of "The Etiology of Alcoholism" is its definition of the research paradigm on which Williams' later alcohol studies are based. While the first biochemical research of vitamins is centred around the study of single-celled organisms, it is here suggested that "the resemblances between the metabolic patterns of rats and humans are marked, and it is for this reason that so much has been learned about human nutrition by studying the nutrition of rats" (Williams, 1947e, p. 570).[216] This reflects the knowledge gained through Williams' own biochemical research with mammals following his relocation to Austin.[217] This equivocation of the metabolisms of humans and rats justifies the use of mammalian subjects to explore the aetiology of alcoholism in humans. The alterations to and individualities of rat metabolism regarding alcoholism, and the potential for its nutritional therapy, are central to Roger Williams' own research on Biochemical Individuality. In "The Etiology of Alcoholism", the extrapolations of this research to human applications are qualified.

7.3 "Alcoholics and Metabolism"[218]

Following the announcement of his increased interest in the research of alcoholism in a technical publication, Williams continues to promote his work and theories in popular science magazines.[219] As is customary for popular science,

---

215 See Section 9.3. *Nutrition and Alcoholism*.
216 See Section 4.2. "The Vitamine Requirement of Yeast".
217 See Section 5.2. The Vitamin Content of Tissues.
218 (Williams, 1948c) To improve the clarity of this section-title and avoid confusion, the citation for this quote is presented in this footnote, deviating from the citation style used in the rest of this dissertation.
219 See Section 7.2. "The Etiology of Alcoholism".

these articles do not provide any references for the information they discuss. With most examples of individuality contained therein already presented in "Alcoholics and Metabolism", they do not offer new information to those having already read Williams' prior publications. They are of increased interest here because Williams speaks of individuality not only as a theory or concept, but as manifest fact in the articles discussed below. What was previously presented as an indication of BI now is presented as proof thereof. Formulations such as "the genes that control our inheritance are capable of so many different combinations that no two persons are exactly alike" (Williams, 1948c, p. 50), and "if the metabolism of individual people is distinctive, then there must be differences in the chemical processes taking place within them" indicate this new assuredness (Williams, 1948c, p. 51). Williams later states that he first became interested in alcoholism as an issue of individuality, with other research aspects resulting thereafter (Hodge and Williams, 1980d).

Within the article "Alcoholics and Metabolism", published in *Scientific American,* Williams publishes multiple diagrams and maps indicating the process of alcohol metabolism and the spread of alcoholics within the United States. Additionally, the social ramifications of alcoholism are indicated by graphs comparing the marital status of the "general population" and "inebriates" (Williams, 1948c, p. 53). The use of such maps and depictions remains unique to this article, though other styles of illustration are utilised in later works.[220] The aspect of environmental influences on the nutritional needs of humans, as well as psychological modulation of metabolism discussed previously, is reaffirmed.[221]

A final noteworthy aspect of "Alcoholics and Metabolism" is an allusion to his own research on "inborn differences" being performed at DBUT, which indicate "distinctive metabolic traits" are evidenced through "careful analysis of (...) body fluids" (Williams, 1948c, p. 51). Such research, using similar phrasing, is released the following year (Williams et al., 1949b, p. 265). This public advertisement for the research of DBUT, and the self-stylisation as an expert regarding individuality, is new in comparison to the previously discussed works published by Williams. Following this publication, Williams increasingly begins to cite his own work and research regarding individuality. Conversely, *The Human Frontier* merely contains a single self-citation (Williams, 1946a, p. 70). As Williams had not produced own research on BI prior to *The Human Frontier*'s publication, and

---

220   See Sections 8.1. The Metabolic Individualities of Rats, 9.4. Biochemical Institute Studies, 10.2. Signatures, and 11.5. Normal Young Men.
221   See Section 7.2. "The Etiology of Alcoholism".

the book dedicates little space to the topics he had previously researched, this lack of self-citation can reasonably be explained. *The Human Frontier* is often cited as a source and comprehensive overview of examples of Biochemical Individuality following its publication. In Williams' research papers on vitamins, a field in which he is celebrated as the discoverer of pantothenic acid and undoubtedly considered an expert, his own work often appears in the bibliographies of his publications.

Having amassed a certain understanding of the individualities of human beings, Williams seems more confident in his knowledge on the subject. The announcement of the promise of his own research underlines the validity of this hypothesis. Williams increasingly begins to publish articles with allusions to his own ongoing research and promoting the work of DBUT following "Alcoholics and Metabolism". Exemplary here is the article "Biochemical Approach to Individuality", appearing in *Science* (Williams, 1948d). Presenting no novel information, its sole purpose is to further promote the concepts presented in *The Human Frontier* and his own research thereupon. Repeating the "unparalleled importance" of such study, Williams shows confidence in this new scientific field (Williams, 1948d, p. 459).

In his later career, Williams is an outspoken defender and promoter of an individualised approach to medicine and nutrition. He is often stylised as "a foremost authority on the science of nutrition" and seeks to present his research and knowledge in a simple and easily understandable way (Williams, 1962, Cover). "Alcoholics and Metabolism" doubtlessly represents such a form of popular scientific publication, yet is much less direct in its stylisation. This outward form of allusion to Williams' research is conspicuous as a precursor to the later common explicit placement and self-citation of his own publications.

7.4 *An Introduction to Biochemistry*, Second Edition

In October of 1948, the second edition of *An Introduction to Biochemistry* is published by D. van Nostrand Company Inc. (Williams and Beerstecher Jr., 1948a).[222] The university textbook represents a complete overhaul of the first edition, which was released 17 years beforehand (Williams and Beerstecher Jr., 1931).[223] This new edition appears with drastically altered prerequisites when

---

222   See Footnote 64.
223   This becomes most clear when regarding the reference sections, which are provided at the end of each chapter (Williams and Beerstecher Jr., 1948a, pp. 34, 72, 73, 94, 95, 132, 160, 161, 195, 211, 218, 228, 286, 287, 335, 350, 399, 412, 445, 466, 477, 489,

compared to its predecessor. Williams and Beerstecher (1948a, p. iii) summarise the scientific context in the preface of their work:

> When the first edition of this book was published in 1931, the status of biochemistry as a separate field of science was much more tenuous than it is today. It depended for its justification largely upon its application either to medicine or to agriculture.
>
> Now, biochemistry has reached a more mature and less dependent stage. Statistics show (...) that in recent years as many doctorate degrees have been awarded in this field as in any branch of science save chemistry. There is a growing tendency to consider biochemistry – the chemistry of living things – as a field of learning well worth cultivation in itself and to accord it a place in academic as distinct from professional curricula. (Williams and Beerstecher Jr., 1948a, p. iii)

Under the described circumstances, it is unsurprising that the textbook does not limit itself to biochemical basics. Biochemistry having been established as an autonomous field of science, the exploration of its boundaries and the possibilities it brings is a natural next step. While the largest part of the book does focus on primary knowledge of biochemical processes, multiple sections and allusions emphasise the importance of new research and the applications thereof. Included in these are a plethora of references to BI, though they are not all equal in their directness and relevance. Some, such as the variation in amino acid requirements between different strains of Bacteria on the one hand (p. 207), and the B-Vitamin-requirements of Protozoa on the other (p. 220), indicate an individuality without providing further information or evidence towards these differences (Williams and Beerstecher Jr., 1948a). The fact, that "some organisms have very unusual requirements and probably unusual chemical compositions as well, though the latter point has not been adequately investigated", addresses BI more directly (Williams and Beerstecher Jr., 1948a, p. 223). The most direct reference to BI appears in the section titled "Idiosyncrasies" of Chapter XII: "The Nutritional Requirements of Mammals":

> We cannot take space more than to indicate that individual peculiarities exist and that food that may be entirely satisfactory for one individual may poison another. This fact will be mentioned in later discussions.

509, 520, 530, 561, 591, 592, 624, 625, 647, 683, 697). Merely five references in the entire book pertain to works that can be clearly dated prior to the publishing date of the first edition (Williams and Beerstecher Jr., 1948a, pp. 195, 211, 445, 592). Two of these works published earlier, *Chemical Embryology* and *The Principles of Plant Biochemistry*, were both published in 1931, though it is not possible to ascertain whether or not they were published before May. At the time of writing, no first edition is available for direct comparison.

> There are many current notions concerning the value or lack of value of various combinations of foods. These notions as a rule have very little scientific foundation (…) This does not preclude the possibility that some of these ideas may some day be proved current. It is dangerous, however, for laymen to pin too much faith on assertions which are not backed by scientific experiment. The human machine is infinitely more complex than a motor car and experimentation is the only safe way to find out its needs. (Williams and Beerstecher Jr., 1948a, pp. 281–282)

Though a comparatively short section within the 736 pages of *An Introduction to Biochemistry*, this excerpt contains multiple important concepts. While Williams often criticises the lack of study toward BI, this constitutes the first reference to the possibly far-reaching ramifications of this shortage of study. This textbook is aimed at an audience of young undergraduate students who are, by definition, less versed in scientific research and the significance of evidence than Williams' colleagues are. Such warnings of the possible consequences of basing too much on scientific theory without practical evidence of its validity is reflected in all of Williams' publications on BI. Though presenting a theoretical concept, Williams seeks to search and provide evidence for its practical legitimacy, alternatively calling for further research in order to prove it, as is typical for the scientific method. Befitting its nature as a workbook, the textbook is more explanatory than his scientific publications and uses more palpable and day-to-day models, much like the popular science articles discussed previously.[224] An example thereof can be found in the section on "Metabolic Rate" in Chapter XXVII: "Intermediate Lipide Metabolism":

> The natural tendency of individuals to carry on combustions is greater than for others. This shows itself when the basal metabolism is determined. (…) If two individuals, such as those used in the illustration above, were to start at the same weight and eat exactly the same food their weights would not stay the same unless their total metabolisms were also the same. If the metabolism of one were one per cent above the food consumption level and the other one per cent below the disparity of weight at the end of ten years would be about 70 lbs., as above. (…) When basal metabolisms are measured clinically variations of 10 to 15 per cent are regarded as within the 'normal' range. (Williams and Beerstecher Jr., 1948a, pp. 643–644)

The example utilised by Beerstecher and Williams to explain the complex concept of metabolic rate is simple, yet striking. In contrast to an example pertaining to laboratory animals or the complex measurements thereof, the image of

---

224 See Chapter 6. *The Human Frontier* and Sections 7.1. "Biochemical Individuality and Its Implications", 7.2. "The Etiology of Alcoholism", and 7.3. "Alcoholics and Metabolism".

humans gaining or losing weight is straightforward. Williams (1956a, pp. 158-159) later uses a similar example in *Biochemical Individuality: The Basis for the Genetotrophic Concept*, with relation to "self-selection". Finally, Williams recommends his own *The Human Frontier* as further reading on "special topics" at the end of Chapter XII, inviting a new generation of scientists to explore his concepts (Williams and Beerstecher, 1948a, p. 287). Having previously addressed established academics and the general public, Williams here presents his theories to budding scientists.

The nature of Williams' approach to *An Introduction to Biochemistry* is not the only remarkable aspect of the publication. The section on "Therapeutic Use of Food Factors" in a Chapter on "The Nutritional Requirements of Mammals" poses a theory very similar to the "Genetotrophic Approach" presented in *Biochemical Individuality: The Basis for the Genetotrophic Concept* (Williams, 1956a, p. 166).

> In the application of the knowledge gained to human beings, sight should not be lost of the fact that the human race is not inbred and uniformity of performance is not to be expected. (…) [W]e should expect in human beings a wide variation in the requirements for any given vitamin due to inborn differences in metabolism, or to induced differences attributable to previous history.
> This means that some individuals probably have much higher requirements for specific vitamins (or minerals or amino acids) than do others. It may easily be that diseased conditions of a mild character, or of long standing, may increase the need for a specific vitamin and that a single individual out of a group on the same diet may be suffering from a deficiency, whereas the others are entirely free from such a lack. (Williams and Beerstecher Jr., 1948b, pp. 283-284)

When compared with the definition of the Genetotrophic Principle, many similarities become apparent.[225] Both sources speak of inborn individuality of nutritional needs as well as the consequences which follow a deficiency that appears

---

225 Williams defines the Genetotrophic Principle in the following manner in *Biochemical Individuality*: "every individual organism that has a distinctive genetic background has distinctive nutritional needs which must be met for optimal well-being. (…) If, during adulthood, the individual (…) fails to meet his particular nutritional needs, he becomes deficient, and this deficiency may contribute to all manner of disease and disease susceptibility. (…) As the individual ages, some of his organs and tissues fail earlier than others because, in accordance with their genetic pattern, they have special characteristics or weaknesses. Those weaknesses may involve unusually high nutritional needs for specific substances which are not provided adequately by the environment in which the organ or tissue resides" (Williams, 1956b, p. 167).

due to a neglect of these needs. The principal difference remains the conclusions drawn from these theories. *An Introduction to Biochemistry* remains vague on whether diseases may be caused by such a deficiency, stating that pre-existing diseases may exacerbate the need for a specific nutritional compound (Williams and Beerstecher Jr., 1948a, pp. 283–284). This could stem from the lack of overwhelming evidence at the time of publication. "The Genetotrophic Approach", however, dedicates 11 pages to the meaning and ramifications of this principle (Williams, 1956b). According to Williams (1956a, p. 173), "the genetotrophic idea was first conceived in connection with the experimental work on alcohol consumption by rats"; this research was published after *An Introduction to Biochemistry*.[226] Similarly, the works referenced in the section dealing with the same topic in *Biochemical Individuality* are all published following the release of the textbook.[227] This, in addition to the evidence for the origins of GC provided by Williams, clearly demarcates this section of "Therapeutic Use of Food Factors" as the first broader presentation of a basic form of GC, as opposed to the schematic thoughts, which are later reflected in GC, discussed previously (Williams and Beerstecher Jr., 1948b, pp. 283–284).[228] Though the ideas presented in *An Introduction to Biochemistry* do not identically match the definition of GC, they contain multiple important elements of the latter theory.

An aspect of *An Introduction to Biochemistry* which is difficult to determine is the understanding of outside influences thereupon. Though "monographs and reviews" (p. 33) depicting the overall scientific standard of the time are cited, "journal articles (…) [are] not, in general, (…) referred to individually" (Williams and Beerstecher Jr., 1948a, p. 33). The references to assorted reviews are, therefore, the only indication of possibly influential works and scientists the reader is given. Merely three sources cited in *An Introduction to Biochemistry* reappear in *Biochemical Individuality*, though none are cited for the first time by Williams

---

226 See Section 8.1. The Metabolic Individualities of Rats.
227 The following sources are referred to on page 173 and listed on page 176 of *Biochemical Individuality: The Basis for the Genetotrophic Concept*: "[12]Roger J. Williams, L. Joe Berry, and Ernest Beerstecher, Jr., *Arch. Biochem.*, **23**, 275–290 (1949)., [13] Ernest Beerstecher, Jr., Janet G. Reed, William Duane Brown, and L. Joe Berry, *Univ. Texas Publ.*, **5109**, 115–138 (1951).,[14]Janet G. Reed, *Univ. Texas Publ.*, **5109**, 144–150 (1951).,[15]Janet G. Reed, *Univ. Texas Publ.*, **5109**, 139–143 (1951)." (Williams, 1956a, pp. 173, 176). As shown by their publication dates, all works appeared after October 1948, when *An Introduction to Biochemistry* was released (Beerstecher Jr. et al., 1951; Reed, 1951a, 1951b; Williams and Beerstecher Jr., 1948a).
228 See Section 5.3. Practical Applications of Vitamin Research.

in the textbook.²²⁹ Though the length of the book and the citation of relevant reviews and monographs indicate intensive study of sources and research results, it does not appear as if outside influences significantly played into novel aspects of BI presented in *An Introduction to Biochemistry* discussed above.

An Introduction to Biochemistry presents the varied aspects of BI to students of biochemistry. Furthermore, it is the first publication to present the preliminary thought-process behind GC. Though further vague allusions to BI can be found in *An Introduction to Biochemistry*, no further discussion of these will be presented here.²³⁰ Previously discussed aspects of individuality, such as the different growth patterns of yeast strains, also reappear and, as they have been extensively discussed elsewhere, will not be expanded upon at this time.²³¹

7.5 Anti-Communist Sentiment

Though the previous sections of this dissertation have largely highlighted the purely scientific aspects of Williams' publications and research, the political aspects of his work become clear when regarding its historical context. Williams' search for individuality, away from "the hypothetical average man", begins to gather momentum in the years following the Second World War (Williams, 1948b, p. 10). The anti-communist sentiment, which dominated American public opinion throughout the Cold War, is reflected most obviously in one of Williams' publications from 1948. Though the radical anti-communism of the early Cold War, so-called McCarthyism, is usually described as a phenomenon of the 1950s, voices and opinions opposed to the political system of the Union of Socialist Soviet Republics (USSR) became louder throughout the second half of the 1940s (Powers, 1998, Chap. 8).²³²

---

229 See Tab. 2: Articles Cited by *Biochemical Individuality* Appearing in Prior Publications.
230 The two most direct and relevant references and theories posed in the textbook have been deliberated on above. Vaguer allusions, or those later less relevant to Biochemical Individuality, are not referenced or discussed for the sake of brevity. A total of 24 references to BI, or aspects thereof, can be found within *An Introduction to Biochemistry* (Williams and Beerstecher Jr., 1948a).
231 See Sections 4.2. "The Vitamine Requirement of Yeast" and 5.1. Individuality of Yeast Strains.
232 McCarthyism is the term used to describe "the use of unscrupulous methods of investigation against supposed security risks and the creation of an atmosphere of fear and suspicion" (Mervin, 2018). It refers to Senator Joseph McCarthy of Wisconsin, who was outspoken about the risk of communist infiltration of the

In "Why People Are Different", Williams voices how his concepts are diametrically opposed to the fundamental ideas of a communist system. "Our very liberty", so Williams, "is based upon our differences" (Williams, 1948b, p. 10). The appreciation and accommodation of these differences is, by logical conclusion, central to a free society. A communist state, however, relies on the uniform treatment of all of its citizens and therefore fails to accommodate for individual needs. "This is a fundamental reason why a Communist regime cannot last" (Williams, 1948b, p. 10). The remainder of this short article contains an abridgement of the research and examples discussed previously. Williams politicises his theory with this statement published in the IBM-Magazine *THINK*.[233] As previously described, Williams suggests the possible grandeur of the results to come from the study of humanics, including world peace and the solution of the greatest societal problems, in *The Human Frontier*.[234] Here, his statement sets out the political consequences of his research: communism cannot adequately provide for the individuality of human beings. Williams makes a similar statement, describing how human's individuality drives them toward freedom and liberty, explicitly naming "democracy and equal opportunity" as the precursors to the ideal development of every individual, in a talk later that same year (Williams, 1950b, p. 270). His criticism becomes more profound in his accusation that communists "seek to shape their science to fit their purposes, deny the facts of heredity and exaggerate the ability of human beings to fit into the pattern of the state" (Williams, 1948b, p. 10).

Williams does not, however, make an ideological difference in is critique of the neglect for study of individuals; here democracies as well as communist states are considered to be seriously lacking. It follows that Williams himself had no sympathies for the Soviet Union and its political ideology, and no indication of any communist affiliations or sympathies can be found in his papers. These comments further emphasise his conviction and deep-rooted belief in the study of humanics as a solution to the problems of society in general, his own brand of political ideology.

Badash (2000) describes the effect of the anti-communist zeitgeist of McCarthyism upon the scientific world and the increasing politicisation of

---

United States government and its institutions. He is seen as responsible for the heated anti-communist sentiment of the early Cold War. For further information, see Michaels (2017).

233 See Footnote 140.
234 See Section 6.1. "Humanity Must Understand Itself".

science and its researchers in his paper "Science and McCarthyism".[235] With growing societal anti-communist sentiment, a similar development can be observed within the scientific community. Williams' open criticism of communist ideology fits well into this depiction of post-war scientific publication. The vilification and consequent alienation of those inclined toward the communist ideal affected teachers, scholars, and scientists alike (Badash, 2000, p. 57). With a team of 17 scientists working under Williams in 1950, various grants are necessary to fund the work of DBUT (Beerstecher Jr., 1950). Receiving grants from military and government agencies was greatly eased if there could be no doubt as to the political affiliations of the researchers involved.[236] Working under the support of various government and military funding, the promotion of Williams' concepts as anti-communist is, therefore, not only an indication of the political nature of BI, but could be an attempt to increase the likelihood of its acceptance in an increasingly anti-communist society (Williams, 1949a; Rogers and Williams, 1954).

7.6 Symposium and Society

Williams often speaks of his theories at symposia to establish and discuss his ideas within the scientific community. In October of 1948,[237] he is invited to speak at *The Second International Symposium on Feelings and Emotions*, sponsored by *The Loyal Order of Moose*.[238] In Williams' terms, the symposium is "dedicated to the better understanding of emotions and feelings in recognition of the importance of these emotions in building human happiness" (Williams, 1950b, p. 268). His talk discusses the intricacies of the individuality exhibited by children and the subsequent failings of the American educational system. The subject matter of his contribution to the symposium heavily leans on Chapter X of *The Human Frontier*, "Humanics and Education" (Williams, 1946a, pp. 185–197).

---

235 See Footnote 141.
236 See Section 6.2. Audience.
237 The publication *Feelings and Emotions*, which contains the written accounts of all talks held at the Second International Symposium on Feelings and Emotions, appeared in 1950. Therefore, the year provided by its citation and the date of the actual presentation do not coincide.
238 The Loyal Order of Moose is a fraternal organisation, giving itself the title "The Family Fraternity". Today, it supports children in need with schooling and community and runs a retirement community. For further information, see Loyal Order of Moose (2024).

The emphasis in this presentation, aside from the Biochemical Individuality that is the basis thereof, is on the importance of physiological individuality. Although the subject matter of humanics has been usurped by BI at this point, the holistic approach of humanics is still entrenched within it.[239] The call for the further research of physiological individuality additionally confirms Williams' move away from humanics, instead highlighting the importance of a more specialised research of individuality in its respective fields. The call for increased personalisation of education remains an important aspect of BI for Williams, culminating in the publication of multiple articles and a book in his later career.[240]

On the 10th of December 1949, Roger Williams is invited to speak in front of *The Philosophical Society of Texas*, a society founded in 1837 "for the collection and diffusion of knowledge" (The Philosophical Society of Texas, 1949, p. 2). Williams, a member of said society, speaks of the theories discussed in the previous sections of this chapter for 84 guests at the Texas Federated Women's Club at Austin (The Philosophical Society of Texas, 1949, p. 3). The protocol of this meeting not only includes the entirety of Williams' talk, but additionally records his introduction by McGruder Ellis Sadler (1896–1966).[241] The contents of Williams' presentation reiterate the topics discussed in this chapter and presents no evidence of changes to BI. Sadler's introduction, however, offers an insight into his standing within society as whole, outside of the scientific world. As the president of a rival university and expert on religion and religious education, Sadler's view on Roger Williams doubtlessly reflects his societal regard apart from his reputation in the scientific world.

> Our speaker for this evening is a scientist who came from Oregon to Texas some ten years ago. He had won recognition for his work in organic chemistry and in biochemistry. Since he arrived at The University of Texas he has developed here a center for research and graduate education which is favorably known throughout the country and the rest of the civilized world. While he has continued his work in biochemistry, he has also developed interests in broader aspects of science and in the application of scientific methods to important human problems. Many of you have read his book, *The*

---

239 See Section 7.1. "Biochemical Individuality and Its Implications".
240 These include *Rethinking Education: The Coming Age of Enlightenment* (Williams, 1986), "Forty Ways To Be Dumb" (Williams, 1957b), "Individuality and Education" (Williams, 1957c), *Free and Unequal* (Williams, 1953a, Chaps. 7 and 11), and "A Flaw in Medical Education" (Williams, 1972).
241 McGruder Ellis Sadler was the president of Texas Christian University and a member of *The Philosophical Society of Texas* in 1949 (The Philosophical Society of Texas, 1949, p. 38). For more information on his life and work, see Procter (2023).

Human Frontier. Tonight he will deal with some problems in the light of the science of today, but in a spirit that would have delighted the founding fathers of our Society. (The Philosophical Society of Texas, 1949, p. 4)

These opening remarks indicate high regard for Williams and his research, though they do not come unexpectedly in light of the occasion on which they are uttered. Williams' expertise in the field of biochemistry is undisputed, therefore the application of his scientific knowledge onto other fields seems to have been taken favourably by those outside of strict academic sociology, whose reviews of *The Human Frontier* are previously less than favourable.[242] Though Williams' authorship of multiple textbooks indicates a certain expertise within university education, Sadler's words further draw attention to this fact, suggesting an international regard for his teaching.

These words indicate Roger Williams' standing at the time of his talk, however his membership in *The Philosophical Society of Texas*, and fact that he delivers such a presentation before its members, is noteworthy in itself. As highlighted by the records of this meeting, "membership [to the society] is by invitation" only (The Philosophical Society of Texas, 1949, p. 2). Counting merely 143 members in 1949, this society appears to be of an exclusive nature. Williams' invitation to join this group of academics is therefore further indication of his high societal regard at the time (The Philosophical Society of Texas, 1949, pp. 34–39). It must be mentioned that William Lockhart Clayton (1880–1966),[243] the elder brother and business partner of one of Williams' greatest financial supporters and friends Benjamin Clayton (1882–1978),[244] is also listed as a member of the society in 1949. The connection between Williams and Clayton may have played a role in his invitation to the society, though this is merely conjecture. Regardless of this possible link, Williams' involvement in and presentation before *The Philosophical Society of Texas* underlines his increased bid for publicity regarding his theories of BI and GC, as it largely comprises dignitaries outside of the scientific realm.

The title and introduction of his talk "Shall WE Pioneer Too?" are reminiscent of the sentiment of "frontiers" discussed in relation to *The Human Frontier*.[245] Once again, Williams professes the potential for adventure and grandiose

---

242 See Section 6.6. Reviews of The Human Frontier.
243 William Lockhart Clayton was a cotton merchant and businessman in Texas, founding the company Anderson, Clayton and Company in 1905. For further information, see Tinsley (2023).
244 See Section 3.2. Academic Networks.
245 See Sections 6.1. "Humanity Must Understand Itself" and 6.2. Audience.

discoveries, likening the research of human individuality to the pioneering travels of Meriwether Lewis (1774–1809) and William Clark (1770–1838) in the Northwest Territory (Williams, 1949b, p. 7).[246] Recollecting the "spirit of the [Texas Philosophical Society] founders", Williams describes his research on individuality as the continuation of pioneering work and "press[ing] forward to new goals" (Williams, 1949b, p. 6). Quoting Francis Bacon (1561–1626), as did the founders of the society, Williams presents his research as part of the maxim "knowledge is power" (Williams, 1949b, p. 5).[247] This phrase is repeated on three other occasions within the presentation, further underlining its importance (Williams, 1949b, pp. 11, 20, 22). Williams' propensity toward such grandiose descriptions for the potential of his research becomes evident here. In the search for support for his theories and research, it will doubtlessly have functioned well in capturing an audience's attention, especially as the addressees are non-specialists. Such animated outward advertisement of his own and other research remains an important part of Williams' later popular science work, and further underlines the hypothesis that a central aspect of his research was to promote the importance and relevance of (bio)chemistry to society at large.[248] This is similarly highlighted by the multitude of letters which Williams receives thanking him for copies of his manuscript. Williams sent such copies to a large variety of people, including academics, chemists, politicians,[249] religious leaders, doctors, and judges in order to further promote his ideas (Barnard, 1950; Bogert, 1950; Connally, 1950; Cullum, 1950, 1950; Fifield Jr., 1950; Fulbright, 1950; Hart, 1950; Horton, 1950; Hutchins, 1950; Johnson, 1950; Shivers, 1950; Wynne, 1950; Yerkes, 1950). As this talk substantially presents the information discussed in connection with *The Human Frontier*, its further contents will not be further deliberated on here.[250]

---

246 The Lewis-and-Clark-Expedition refers to the exploration of the Northwest Territory of the United States by Meriwether Lewis and William Clark from 1804 to 1806. For further information, see Lewis et al. (1997).
247 Francis Bacon was a British intellectual, lawyer, academic, statesman, and philosopher. For further information on his life and work, see Abbott (2013).
248 See Section 5.4. *What to Do About Vitamins*.
249 Including multiple U.S. senators, the governor of Texas, and Lyndon B. Johnson.
250 See Chapter 6. *The Human Frontier*.

```
E. TYDINGS, MD., CHAIRMAN
IA.        STYLES BRIDGES, N. H.
"          CHAN GURNEY, S. DAK.
           LEVERETT SALTONSTALL, MASS.
IX.        WAYNE MORSE, OREG.
L.         WILLIAM F. KNOWLAND, CALIF.
           HARRY P. CAIN, WASH.

NELSON TRIBBY, CLERK
```

United States Senate
COMMITTEE ON ARMED SERVICES

June 16, 1950

Dear Dr. Williams:

       I have read your address given before the Philosophical Society of Texas with a great deal of interest and am indebted to you for sending me the copy.

       I noted with particular interest your reference to political writers. While I realize this was only one example of many you might have chosen, It caught my eye because it concerned a group with whom I come in contact almost daily.

       I am glad you mentioned the subject in connection with the National Science Foundation and would appreciate hearing from you whenever you might have further suggestions in this respect.

       With best regards,

Sincerely,

Lyndon B. Johnson

Dr. Roger J. Williams
Professor of Chemistry
The University of Texas
Austin, Texas

Fig. 12: Letter by Lyndon B. Johnson to Williams on 16.06.1950 indicating his interest in Williams' address to the Philosophical Society of Texas; Dolph Briscoe Center for American History, The University of Texas at Austin, Roger Williams Papers, Box 88-087/41, Folder: LETTERS: RE "SHALL WE PIONEER TOO?"

7.7 Biochemical Individuality in 1949

Every individual human being differs from his or her peers in multiple aspects. The effect of alcohol on individual human beings is no exception, much as the susceptibility to alcoholism varies within a population. Genetic and environmental aspects affect an individual's vulnerability to chronic alcohol consumption, with nutrition as an important aspect. Metabolic individualities, based on genetic variation, may predispose toward the development of alcoholism, though the disease is of multifactorial aetiology. An approach to medical and psychological research based on the principles of humanics could help us understand a multitude of diseases, including alcoholism.

7.8 Conclusion

Following the publication of *The Human Frontier*, Williams and his colleagues at The Department of Biochemistry at The University of Texas at Austin begin to research the theories posed therein. With alcoholism at the forefront of their studies, the results of their research favourably promote the truthfulness of Williams' concepts. With the conjugate term Biochemical Individuality appearing for the first time in 1947, Williams promotes his thoughts and theories through popular science magazines, his university textbook, and a talk in front of *The Philosophical Society of Texas*, which additionally indicates his growing societal regard. The political tenor of the time, namely anti-communist sentiment and the growing regard for scientists in the United States, is an essential factor in the developments of the post-war years. Concrete evidence for the distinctiveness of individuals in their propensity toward alcoholism lead to the publication of the concept of genetotrophic disease in the following years. The first notion of the Genetotrophic Principle comes to light in the second edition of *An Introduction to Biochemistry*. This concept is later at the centre of Biochemical Individuality and is the overarching theme of the research of the following years.w

8 Genetotrophic Disease

While most of the publications immediately following *The Human Frontier* are more or less abridgements of the aforementioned book, research on the disparities between individuals is underway at The Department of Biochemistry at The University of Texas at Austin in the years following its publication. Having called the scientific community to arms in *The Human Frontier*, Williams is able to publish the first conclusive results from his institute alongside his colleagues in 1949. As announced and justified in "The Etiology of Alcoholism", the study of rats and their behaviour provides the basis for human extrapolation and is applied to Williams' theorems on alcoholism (Williams, 1947e, p. 570).[251] This study of alcoholism is of greatest importance, as it lays the foundation for the Genetotrophic Principle which later dominates Williams' career (Williams, 1954d, p. 802).

8.1 The Metabolic Individualities of Rats

Williams' critique of the predominant ideas regarding "the hypothetical average man" is essential to the hypotheses posed in *The Human Frontier*, therefore Williams prioritises finding scientific proof of this central assumption over further-reaching analyses. The first object of research discussed in an article entitled "Individual Metabolic Patterns, Alcoholism, Genetotrophic Diseases" is to prove the existence of a "metabolic personality"; the variation of metabolism from individual to individual (Williams et al., 1949b, p. 265). Using polar coordinates, Williams et al. (1949b, p. 266) present a direct comparison of the differences between two supposed metabolic personalities.[252] One depicts the chemical makeup of the body fluids of an aforementioned "average" individual and the other an approximation of a "typical" real human being. Though a comparison between an "average" and a "typical" individual seems ironic, the depiction of these sun-like diagrams effectively conveys Williams' central message: individuals are not as similar as medicine, science, and society has made them out to be. Though this comparison is made on the basis of the study of a "real individual"

---

251 See Section 7.2. "The Etiology of Alcoholism".
252 The term "polar coordinates" here refers to sun-like diagrams, in which the length of each "ray" depicts the concentration of certain amino acids in bodily fluids or the strength of a physiological reaction to a stimulus. For further information on polar coordinates, see Clapham and Nicholson (2009) and Cammack et al. (2006).

over the course of "several months", Williams and his colleagues do not specify of what species or descent this individual is (Williams et al., 1949b, p. 265). It may stem from the "preliminary study of the metabolic patterns of a considerable number of patients of the Austin State Hospital" mentioned later in the article (Williams et al., 1949b, p. 271).

Having indicated the existence of Biochemical Individuality in the analysis of body fluids, the understanding it offers is applied to the issue of alcoholism. The consumption habits of individual rats and mice, when allowed to choose between simple water and a solution of 10 % alcohol, is observed and recorded. These individual animals can all clearly be allocated to one of five strains and "each strain of animals appears distinctive with respect to its alcohol consumption" (Williams et al., 1949b, p. 267). This mode of study, a comparison between the behaviours of different strains of the same species, bears many resemblances to Williams' first piece of published research.[253] As in his dissertation, individualities between such strains become evident through the results of his studies. From this, Williams et al. (1949b, p. 267) extrapolate with certainty that behaviour in alcohol consumption, and therefore the predisposition towards alcoholism, is genetically predetermined. Furthermore, the researchers find themselves able to "abolish with surprising regularity the appetite of rats and mice for alcohol" through "somewhat unusual dietary considerations" (Williams et al., 1949b, p. 268). These dietary considerations relate to the addition of ten B vitamins, "the antipernicious anemia vitamin",[254] vitamin A and/or linseed oil to the diet offered to the test subjects (Williams et al., 1949b, p. 268). Following this method, "an appetite for alcohol (…) can be abolished in all rats by a use and extension" thereof (Williams et al., 1949b, p. 270).

The results of this research, which are scientifically published in *Archives of Biochemistry* as the third part of the series titled "Biochemical Individuality" (Williams et al., 1949a),[255] lead to an extension of the hypothesis of the aetiology

---

253 See Section 4.2. "The Vitamine Requirement of Yeast".
254 Also referred to as Vitamin B_{12}.
255 The two papers, "Individual Metabolic Patterns, Alcoholism, Genetotrophic Diseases" (Williams et al., 1949b) and "Biochemical Individuality III. Genetotrophic Factors in the Etiology of Alcoholism" (Williams et al., 1949a) respectively appear in June and September of 1949. Both are nearly identical in their content and much of their wording. Where there are differences, these remain minor and do not alter the meaning or contents of the two pieces. They are treated equally and should be seen as such within this dissertation. Where one is quoted, the other could have been utilised to similar effect, the choice was made purely on the basis of superior

of alcoholism, proposed in the *Quarterly Journal of Studies on Alcohol* in 1947 (Williams, 1947e).[256] While "The Etiology of Alcoholism" suggests the possibility of a genetic predisposition toward alcohol, the research presented in "Individual Metabolic Patterns, Alcoholism, Genetotrophic Diseases" allows further specification. It follows, that "Alcoholic craving (…) constitutes a perverted appetite which arises as a result of one or more dietary deficiencies" (Williams et al., 1949b, p. 270). At the root of these deficiencies is not, as could be concluded, "a failure to eat what is regarded as satisfactory food", but more a diet which is deficient according to the specific needs of the individual as predetermined by his or her specific genetic make-up (Williams et al., 1949b, p. 270). The extensive use of refined foods is additionally hypothesised to play a contributing role (Williams et al., 1949a, p. 286). The cause of these individual nutritional needs are postulated to be "partial genetic blocks" (Williams et al., 1949b, p. 270), a genetic concept published three years earlier by Mitchell and Houlahan (1946).[257] These partial genetic blocks are hoped to be circumvented by the scientists at DBUT by a so-called "shotgun therapy", abolishing the alcoholic's compulsion to drink through the administration of a carefully personalised mixture of vitamins and nutrients (Williams et al., 1949b, p. 270). This approach is akin to the concepts of personalisation to which western medicine aspires today still. Williams highlights his previous critique that "commercial vitamin preparations are often inadequate and poorly balanced", and warns that "haphazard administration of 'vitamin pills' and self-medication by uninformed laymen can be expected to be completely ineffective, or even harmful and dangerous" (Williams et al., 1949a,

---

or more concise wording in one of the two sources. The second piece of this series is titled "Biochemical Individuality II: Variation in the Urinary Excretion of Lysine, Threonine, Leucine, and Arginine" and neither includes Roger Williams as an author, nor contains information pertinent to the development of Biochemical Individuality (Thompson and Kirby, 1949). It concludes that humans differ in their excretion of amino acids in their urine. For a discussion of the first publication of this series, see Section 7.1 "Biochemical Individuality and Its Implications".

256 See Section 7.2. "The Etiology of Alcoholism".
257 Williams describes his understanding of partial genetic blocks as follows: "When there is a partial genetic block, the capability of producing a specific enzyme is not entirely lost, but it is impaired to such an extent that there develops an augmented nutritional requirement which may be for a specific vitamin, amino acid, mineral element or other metabolic substance" (Williams et al., 1949b, p. 270). Williams' understanding of these blocks is pertinent here, as this forms part of the basis of his theory. The original publication and the consequent explanation therein is published by Mitchell and Houlahan (1946).

p. 287). The medical abolition of alcohol, he therefore concludes, should, not be oversimplified or be taken out of professional's hands.

The conclusions drawn from these results are subsumed in the concept of genetotrophic disease,[258] which unites them with the concepts of Biochemical Individuality described in the previous chapters.[259]

> The concept of genetotrophic disease (geneto = genetic; trophic = nutritional) (…) is one arising fundamentally from nutritional deficiency which in turn has its basis in a genetically controlled augmented requirement for one or more specific nutritional elements. (Williams et al., 1949b, p. 270)

This genetotrophic aetiology of disease is further assumed to have potential ramifications for diseases of hitherto unknown source. Concrete examples, such as "allergies, mental diseases, cardiovascular disease, arthritis, multiple sclerosis and (…) cancer" (Williams et al., 1949b, p. 270), as well as psychological disorders, for which Williams and his colleagues have begun to collect human samples for research at this time, are also considered to be among these (Williams et al., 1949b, p. 271).

Though concepts akin to the Genetotrophic Principle have previously appeared in other publications, Williams here designates a scientific term to these ideas. This thought of genetotrophic disease, which can be treated and prevented through adequate nutritional supplementation, provides a basis to much later research on nutrition and its value at DBUT. Having studied the structure and biochemistry of vitamins in the first decades of his career, Williams integrates the knowledge gained from his ongoing research on BI with his expertise to formulate the Genetotrophic Concept. This further underlines the hypothesis that BI develops as an accumulation of increasingly in-depth knowledge of human and animal biochemistry on the basis of vitamin research. While the genetotrophic aetiology of disease is proposed here, it does not reflect the theory presented in *Biochemical Individuality* in all aspects. As for the entirety of the theory of BI, Williams' work undergoes further refinement and optimisation in the years building up to 1956.

Williams and his colleagues do not only present the results of their research in 1949, but additionally pose strata of research necessary to realistically implement the systems of treatment they suggest. The exact makeup of the responsible

---

258 Williams confirms that GC arose from the experimental findings of his alcoholism research in a later publication (Williams, 1954d, p. 802).
259 See Chapters 6. *The Human Frontier* and 7. The Scientific Study of Individuals and Alcoholism.

nutritional deficiencies is unclear to the researchers, as is a means for "assessing individual metabolic patterns", to "determine (...) the biochemical roots of his particular difficulty, and, on this basis, to formulate an adequate nutritional treatment" (Williams et al., 1949a, p. 288). Furthermore, a rethinking of hereditary diseases in the medical field is indicated as necessary. With the field of medical genetics making its first strides at this time, Williams et al. (1949b, p. 268) express their disdain for the fatalistic pessimism of the medical profession regarding genetic diseases. In their view, the possibility of genetic malformations finding expression in genetotrophic diseases sheds the light of opportunity upon diseases formerly considered to be incurable. This reference to the lack of confidence in medicine regarding genetic diseases is also voiced in later articles and publications (Williams, 1951c, p. 123, 1951a, p. 15, 1953c, p. 122).

Williams' research is noticeably altered following the publication of the two papers discussed in this section. While articles on the structure and distribution of vitamins in various species dominate Williams' career up to 1946, the number of articles which deal with fundamental biochemical research begins to starkly decline following the publication of *The Human Frontier*.[260] In the three years between the publication of *The Human Frontier* and "Individual Metabolic Patterns, Alcoholism, Genetotrophic Diseases" / "Biochemical Individuality III. Genetotrophic Factors in the Etiology of Alcoholism", Williams regularly appears as co-author in articles on biochemical groundwork studies. As of 1949 this type of "basic" research becomes much more sporadic and research on alcoholism, BI, and GC is a clear focus (Davis, 2003c).

Having presented the first results of their research on alcoholism, Williams et al. (1949b) pose their concept of genetotrophic disease. The investigation of this concept becomes the overarching theme of the following years, with GC being at the centre of Williams' concept of BI. They are the first of his widespread research on the effect of nutrition on disease and confirm the theories postulated in *The Human Frontier*.[261]

---

260 See Fig. 13: Publications by Roger J. Williams (1919 - 1956).
261 See Chapter 6. *The Human Frontier*.

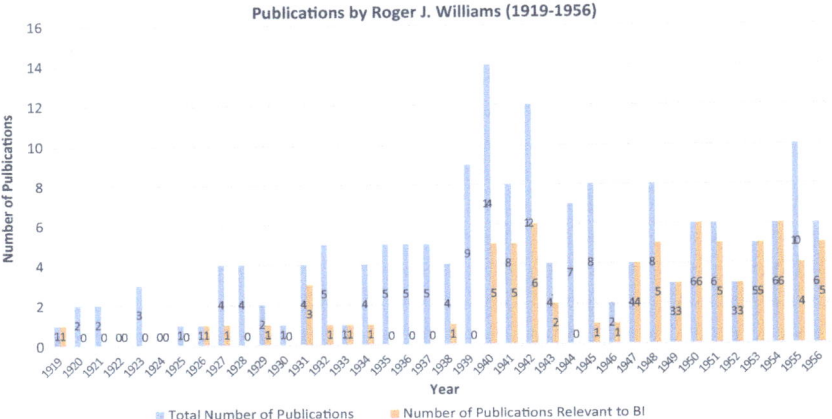

Fig. 13: Publications by Roger J. Williams (1919–1956)

8.2 Collaborative Individuality

Williams' biochemical exploration of vitamins at DBUT is largely characterised by collaborative publications and cooperation in research. His previously discussed publications on various aspects of individuality following the publication of *The Human Frontier*, however, are published by him as the sole author. While this could be taken to signify a lack of collaboration in the research of BI, there are several aspects which indicate the opposite.

On the one hand, Roger Williams is the director of DBUT as of September 1940 (Williams et al., 1966, p. 2). This indicates a high probability of collaboration with scientists working at his own institute. Additionally, in "Biochemical Approach to Individuality", Williams describes how the investigation of individuality constitutes an entire program at DBUT, rather than a single piece of his own research (Williams, 1948d, p. 459). This thought is confirmed by the fact that every piece of research from Austin published as part of the series "Biochemical Individuality" in *Archives of Biochemistry* is authored by various constellations of scientists (Williams, 1947j; Berry and Cain, 1949; Thompson and Kirby, 1949; Williams et al., 1949a, 1950a). It is therefore apparent that, as is the case for the research of vitamins, BI is in certain aspects a collective effort of DBUT.

This partially collaborative nature of BI becomes evident through the genesis of the Genetotrophic Principle. With one exception, every article on GC is published jointly by Roger J. Williams, Ernest Beerstecher Jr. (born 1910),[262] and L. Joe Berry (1910–1987) (Williams et al., 1949b, 1949a, 1950d, 1950c; Williams, 1950c). Ernest Beerstecher Jr., co-author of *The Biochemistry of B Vitamins* and *An Introduction to Biochemistry* (Williams et al., 1950e; Williams and Beerstecher Jr., 1931, 1948a), is Williams' colleague at the time of these publications, working as a research associate at DBUT (Williams et al., 1950c, p. 287). He is later best known for his work in bacteriology and immunochemistry (Beerstecher, Ernest, Jr., 1989). In turn, L. Joe Berry collaborates with Beerstecher and Williams from outside of DBUT, as associate Professor of Biology at Bryn Mawr College in the state of Pennsylvania (Williams et al., 1950c).[263] The cooperation of the three men in developing and promoting GC further emphasises the fact that BI is not purely a result of Williams' own efforts, but that the work and research of his colleagues at DBUT is relevant to the development of BI.[264]

The efforts of other scientists and advisors are acknowledged in all of Williams' publications, though he remains the driving force behind all of these studies throughout the process, and in most of his articles on the topic of BI he remains the sole author. It is additionally worthy of notice that Williams' contribution to the largest publication of results of BI-research from DBUT, *Biochemical Institute Studies IV: Individual Metabolic Patterns and Human Disease: An Exploratory Study Utilizing Predominantly Paper Chromatographic Methods*, is limited to its introduction (Williams et al., 1951). While Williams will doubtlessly have been involved in the processes behind the research, this change from his own basic biochemical research of vitamins in the first three decades of his career to the different form of publications discussed in the following sections is noticeable. In fact, Williams admits his role in this research is one more akin to an observer than an active participant (Williams, 1951d, p. 3). This increased publication on BI without the discussion of his own research continues throughout the first half of the 1950s. It is therefore important to note that many more researchers were

---

262   At the time of writing, it is not possible to ascertain in which year Ernest Beerstecher Jr. died. The date of his death can be placed between 1989 and 1998, because the 1998 edition of *American Men & Women of Science* labels him as deceased, while the 1989 version does not (Beerstecher, Ernest, Jr., 1989, Beerstecher, Ernest, Jr., 1998).
263   Having earned his PhD at the University of Texas at Austin (UT) in 1939, the connection between Berry and DBUT is upheld, culminating in his return to Austin in 1970 (Yee, 2020). Berry is later best known for his work as a medical microbiologist and as an "authority on interactions of infective microorganisms and their hosts" (The Eastern Pennsylvania Branch of the American Society for Microbiology, 1987).
264   See Section 3.2. Academic Networks.

involved in the study of BI, though Williams is doubtlessly the most prominent figure.

8.3 Genetotrophic Promotion

Much like the articles offering an abridged summation of the theories presented in *The Human Frontier*, multiple essentially identical publications follow the Williams' publication of the Genetotrophic Principle.[265] These appear in a variety of journals and magazines and stress different aspects of the theory, according to the superordinate publication in which they appear. Large sections of these articles are duplications of the originals, though they all diverge in certain aspects. An article in *Texas Reports on Biology and Medicine* is an exception, as it is overwhelmingly identical to the original publications (Williams et al., 1950d). These articles effectively promote the idea of GC in different scientific fields related to GC and which could partake in its study and application.

An article appearing in the renowned medical journal *The Lancet* emphasises the possible ramifications of GC for diseases of obscure origin (Williams et al., 1950c). The possible causes of cancer, diabetes, rheumatoid arthritis, disseminated sclerosis,[266] and mental disease are discussed in detail, with genetotrophism posed as one factor in a principally multifactorial aetiology. The term "multifactorial" is of essence here, as the researchers from DBUT do not pose the diseases they discuss as purely genetotrophic; the importance of other factors is often stressed (Williams, 1950c, p. 260; Williams et al., 1950c, p. 6599). Additionally, Williams et al. go into further detail on the possible medical applications compared to articles promoting GC.

A reaction to this article by Lionel S. Penrose (1898–1972) also offers a first glimpse of the reception of GC.[267] Penrose, recognising the similarities of theories to the work of Archibald Garrod (Williams et al., 1950c), takes issue with the portrayal of GC as a new "fundamental concept" (Penrose, 1950, p. 464).[268] Criticising the complexity of GC as "nebulous", Penrose (1950, p. 464) prefers the "clear and really fundamental" theories proposed by Garrod and supposes the authors are unaware of Garrod's research. As has been indicated elsewhere,

---

265 See Section 7.1. "Biochemical Individuality and Its Implications".
266 Also referred to as Multiple Sclerosis (MS).
267 Lionel Sharples Penrose was a well-known and respected human geneticist at University College London. For further information on his life and work, see Laxova (1998).
268 See Chapter 1. Introduction and Section 6.4. "Distinctive Metabolic Traits".

Garrod's work is indeed fundamental to Williams' theories, though Penrose takes these similarities, and possibly the lack of reference to Garrod in this article, as an indication for a lack of ingenuity and ignorance.[269]

The final published article in this form, appearing in *Nutrition Reviews*, elucidates some finer aspects of nutritional genetotrophism not included in other articles (Williams, 1950c). Though Williams is acknowledged as sole author, the article covers the identical topics as those discussed above and similarly duplicates large sections. The collaboration of Beerstecher and Berry in this article can therefore be equally assumed. Such promotion of GC and BI is not, however, limited solely to the articles discussed above. The section written by Ernest Beerstecher as part of the textbook *The Biochemistry of B Vitamins* additionally makes reference to their collaborative research (Williams et al., 1950e, sec. C). Quoting a long passage from *The Human Frontier* for his section on "Inherent Individual Variations in B Vitamin Requirements" (Williams et al., 1950e, pp. 273–274), Beerstecher later makes direct reference to "Biochemical Individuality III. Genetotrophic Factors in the Etiology of Alcoholism", noting the merit of the theory he helped formulate (Williams et al., 1950e, p. 433). Penrose's critique might have had an effect on the formulation of this section, as the thought process behind GC is now described as "not entirely new" (Williams et al., 1950e, p. 433).

With GC first explicitly formulated as a theory in 1949, Williams and his colleagues publish multiple pieces promoting their concept in various articles and a textbook throughout 1950. A further publication, "The Experimental Treatment of Genetotrophic Disease", remains unpublished (Williams et al., 1950b).[270] These promotional activities concluded, the next part of the series "Biochemical Individuality" is published in November of 1950.

8.4 Biochemical Individuality Following the Genetotrophic Principle

The probability of alcohol consumption in rats can be mitigated by the administration of a diverse mixture of vitamins. In all likelihood, the medical and social

---

269 See Section 6.4. "Distinctive Metabolic Traits".
270 A manuscript of "The Experimental Treatment of Genetotrophic Disease" can be found in the Roger Williams Papers, Box 88-087/20a, Folder: The Experimental Treatment of Genetotrophic Disease at the Dolph Briscoe Center for American History of The University of Texas at Austin. No publication under the same title appears in the lists provided by Davis (2003c).

issue of alcoholism could be solved by the identification and consequent nutritional treatment of those individuals whose genetic makeup makes them likely to develop craving for alcohol. Such diseases, arising from a specific nutritional deficiency due to the genetic peculiarities of an individuals, can be described as genetotrophic. Many illnesses contemporarily classified as genetic and untreatable may also be of such origin.

8.5 Conclusion

The behavioural analysis of rats with regard to their alcohol consumption allows Williams, Berry and Beerstecher to formulate their concept of genetotrophic disease. Insight into a possible genetotrophic origin of alcoholism and the study thereof in turn allows extrapolation into a possible treatment for those afflicted with addiction. The collaborative nature of this theory and the entire research of BI is evidenced by the continued joint publication of articles promoting GC. Individuality becomes an increasingly important factor in Williams' research, as evidenced by the relative number of his articles discussing this subject per year. Multiple other diseases of obscure origin, which may be of a genetotrophic nature, are additionally defined and examined to this end. With the lessons learnt from rodent experimentation providing evidence for the relevance of GC, the study of human beings is the next step in the development of BI.

9 Human Individuality

While most medical groundwork research is still based on experimentation with rodent mammals today, the validity of their results for the prediction of human behaviour continues to be controversial (Bracken, 2009; Hackam and Redelmeier, 2006; Shanks et al., 2009). Williams, with an appreciation for the individuality of all organisms, recognises the limited cogency of merely observing rodents when the application of the research is intended for an entirely different species. The confirmation of the genetotrophic aetiology of alcoholism and Biochemical Individuality is, therefore, sought through the analysis of human subjects.[271] This chapter discusses the first results of explicit research on the individuality of human subjects from DBUT up to the publication of *Free and Unequal* in 1953 (Williams, 1953a).

9.1 Biochemical Individuality V

Having found an indication for individuality regarding alcoholism in rats and mice in 1949 (Williams et al., 1949a, p. 265), Williams and his colleagues are able to publish the results of their first biochemical human analysis by the end of 1950.[272] This study, which "explore[s] in a preliminary way various traits of

---

271 See Sections 7.1. "Biochemical Individuality and Its Implications", 7.2. "The Etiology of Alcoholism", 7.3. "Alcoholics and Metabolism", and Chapter 8. Genetotrophic Disease.

272 The completion of the study presented in "Biochemical Individuality. V." in January of the same year could be taken to indicate that this research is conducted simultaneously as the formulation of the Genetotrophic Concept in 1949 (Beerstecher Jr. et al., 1950, p. 27). It does not, however, seem likely that the preliminary results of these studies influenced the development of GC. Though Williams, Beerstecher, and Berry are all co-authors of "Biochemical Individuality. V.", and will therefore have had knowledge of the preliminary results of this study, it is unclear in which timeframe the "4 consecutive weeks" of observation took place (Beerstecher Jr. et al., 1950, p. 30). With "Individual Metabolic Patterns, Alcoholism, Genetotrophic Diseases" and the first formulation of the Genetotrophic Concept therein presented to the *National Academy of Sciences of the United States of America* in April of 1949 (Williams et al., 1949b, p. 265), the theory can be seen to predate the publication of "Biochemical Individuality. V.". Additionally, the theory of genetotrophic diseases is portrayed as pre-existing in the paper itself (Beerstecher Jr. et al., 1950, p. 27). The fifth instalment of "Biochemical Individuality" is therefore treated separately from the origins of GC.

alcoholic and non-alcoholic individuals" (Beerstecher Jr. et al., 1950, p. 27), reveals "common metabolic characteristics" of alcoholics, which non-alcoholic individuals lack (Beerstecher Jr. et al., 1950, p. 37). Over a timespan of four weeks, the researchers at DBUT analyse and compare 62 aspects of saliva, urine, and blood samples of twelve subjects. Of these subjects, eight are non-alcoholic controls and four are members of *Alcoholics Anonymous*.[273] The comparatively small cohort size is deemed acceptable due to the fact that "this is an exploratory study designed to reveal differences which subsequently need to be studied intensively" (Beerstecher Jr. et al., 1950, p. 29). Statistical significance is found for six of the items studied, and they are presented in tables with other statistically insignificant diversions of alcoholic's metabolism (Williams et al., 1950a, p. 39). It is noteworthy that the values obtained for all studied items show clear and in part substantial variation within the control group. This indicates an individuality of all members of the control group, though this fact is not commented on within the article. This fact is, however, picked up and rediscussed in a publication four years later (Williams, 1954 f, p. 196).[274] These results confirm the hypothesis of a possible differentiation between the metabolism of alcoholics and non-alcoholics, providing a basis for more far-reaching biochemical research of alcoholism. This research is already ongoing at the time of publication (Beerstecher Jr. et al., 1950, p. 39).

The data of these further studies are published in May of 1951, using the identical paper chromatographic methods invented and developed by researchers at DBUT (Williams et al., 1950a, p. 28; Kirby Berry et al., 1951).[275] This aspect of methodological and conceptual ingenuity at DBUT is not unique to the research of individuality. As discussed elsewhere, Williams begins to deviate from conventional scientific methods early in his career.[276] DBUT's vitamin research of the 1940s necessitates and brings forth the development of innovative biochemical methods of research and analysis, enabling the institute to discover multiple vitamins (Pennington et al., 1940; Snell et al., 1940a; Williams, 1940, 1941a;

---

273 *Alcoholics Anonymous* is an international non-profit organisation of men and women addicted to alcohol, promoting and supporting the sobriety of its members. It holds regular meetings for its members in order to support them in their goal to abstain from consuming alcohol. For further information, see Alcoholics Anonymous World Services, Inc. (2017).
274 See Chapter 8. Genetotrophic Disease.
275 These paper chromatographic methods are presented as part of *Biochemical Institute Studies IV* in 1951 (Kirby Berry et al., 1951).
276 See Section 4.2. "The Vitamine Requirement of Yeast."

Williams et al., 1941, 1941; Eppright and Williams, 1944; Williams et al., 1944; Hac et al., 1945).[277]

Though other "traditional" laboratory methods are available at this time, the high frequency of the analyses conducted in the course of this study necessitates a less expensive process (Beerstecher Jr. et al., 1950, p. 28). Regarding BI, the "number of special methods (…) not yet (…) described in the scientific literature" offer a financially sound, if less accurate, method of human analysis, "ideally suited to this type of investigation" (Beerstecher Jr. et al., 1950, p. 28). These aspects of resourcefulness and originality regarding research methodology reflect Williams' proclivity toward innovation in all aspects of society following his insights into BI. A reshaping of scientific methodology in the research of new scientific fields is taken in the stride of research and continually refined under Williams' leadership of DBUT. The specific aspects of these methods are presented as part of *Biochemical Institute Studies IV: Individual Metabolic Patterns and Human Disease: An Exploratory Study Utilizing Predominantly Paper Chromatographic Methods* in the following year (Williams et al., 1951).

"Biochemical Individuality V" presents new insights into the metabolism of alcoholics, contrasting these results with the values of non-alcoholic individuals. Research of alcoholism and the peculiarity of associated metabolism continues at DBUT following encouraging results. Roger Williams continues to promote the concepts of GC and BI, giving presentations and publishing articles on various aspects of his theories.

9.2 Of Marbles and Men

A talk entitled "Men and Marbles", held by Roger Williams at an awards luncheon in December of 1950, comprises the portrayal of a model which perfectly encapsulates the concept of BI (Williams, 1951d). Though the talk does not contain novel information on the development of BI, it is discussed here because of the effectiveness of the image it portrays. It impresses the essence of BI upon the reader and finds an effective representation for a concept criticised as too unclear by some of Williams' contemporaries.[278]

While receiving the "Southwest Regional Award" of the *American Chemical Society*, Roger Williams is given the chance to speak before his colleagues at the ceremonial luncheon which follows (Williams, 1951d, p. 1). While accepting this accolade, Williams readily grasps the opportunity to impress the importance and

---

277 See Chapter 5. Vitamin Studies.
278 See Section 8.3. Genetotrophic Promotion.

relevance of BI upon his colleagues (Williams, 1951d, p. 3). Throughout the talk, an image of a "bushel basket" of marbles represents all human beings, while various properties of this assortment of marbles are explored (Williams, 1951d, p. 2). In this thought experiment, Williams invites his listeners to explore the ideal method for the analysis of said marbles, with the object of this research being a complete knowledge and understanding thereof.

The formulation of average values can, it follows, only offer practical information when all marbles are alike in every aspect which is measured. If, say, some marbles are respectively comprised of steel, glass, and plastic, the average value for the content of iron, silicate, and carbon of all marbles would provide values which fail to describe any singular marble accurately (Williams, 1951d, p. 1). In an assortment of red, blue, green, black, white, and colourless marbles, the average colour would most likely be determined to be dirty grey, failing to portray the colour of any marbles in the collection. A similar development can be seen for the average worth of the marbles, if some few are very valuable, and the rest are very cheap (Williams, 1951d, p. 2).

An appropriate and meaningful examination of a heterogenous group, therefore, is only possible through an observation of each individual marble. As this would be impractical and time-consuming, the categorisation of each individual into predefined groups according to the aspects under scrutiny would allow a more accurate picture, while still allowing the calculation of meaningful averages. These "group parameters" can only be defined following the analysis of a small cohort of representative marbles, providing the basic ranges of certain values. The logical conclusion follows that an adequate, yet practical, appreciation of human individuality must be pre-empted by the development of such a system of classification (Williams, 1951d, p. 2).

This portrayal of individuality, the proverbial typical human being a "dirty grey" average, effectively condenses BI into a relatable visualisation of the concept. The imagery is described as "[bringing] the point home with a staggering impact" in a review of the speech by Walter J. Murphy (1899–1959) appearing in *Chemical and Engineering News* on Christmas day of 1950 (Murphy, 1950, p. 4529).[279][280] Murphy's response to Williams' presentation is highly positive, the

---

279 Walter J. Murphy was a chemist and notable figure in the realm of chemical publishing. In December of 1950, he was the editor of *Chemical & Engineering News*, a magazine which he helped to develop and establish. For further information on the life and work of Walter J. Murphy, see "Walter J. Murphy" (1959) and Hallett (1960).
280 *Chemical and Engineering News* is a weekly news magazine published by the American Chemical Society since 1923. For further information, see Chemical & Engineering News (2023).

aforementioned comment confirms the thought that this portrayal more effectively brings across the essence of BI than previous attempts. The bestowal of the *1950 Southwest Award for Outstanding Achievement in Chemistry* is in itself a further indication of the growing esteem of Williams' theories in the scientific community. With BI originally posed as a theoretical scientific possibility of study, Murphy (1950, p. 4529) suggests that DBUT's study of individuality sufficiently demonstrates the practical ramifications and possibilities thereof.[281] The demonstration of a feasibility for the theories, therefore, increases the appreciation for Williams and his research. The presentation "Men and Marbles" offers a new visualisation of Biochemical Individuality, easily understood and applicable to human beings. A review of the speech by Walter J. Murphy provides an indication of the growing regard for Williams' theories within the scientific community and the effectiveness of the aforementioned image.

9.3 *Nutrition and Alcoholism*

Having gained considerable insight into the possible genetotrophic aetiology of alcoholism, Roger Williams publishes his third book in sole authorship, *Nutrition and Alcoholism* (Williams, 1951a). Completed in March of 1951 and appearing the same year (Williams, 1951a, p. X), it bears many similarities to his first popular science book *What To Do About Vitamins* (Williams, 1945a).[282] In line with its presumed audience, namely "alcoholics and their families" (Williams, 1951a, p. 3), the book uses laymen's terms and elucidates the absolute basics of metabolism (Williams, 1951a, chap. 2). It is also published by The University of Oklahoma Press and sells 7545 copies by July of 1953 (University of Oklahoma Press, 1953b). Furthermore, it constitutes a condensation of all knowledge of nutrition and its connection to alcoholism, drawing further parallels to *What to Do About Vitamins*. Though an abridgement of all previous research, *Nutrition and Alcoholism* presents some information not discussed in previous sections of this dissertation.[283] It was later recalled from publication, as Williams feared it may offend members of Alcoholics Anonymous (Williams, 1958; Hodge and Williams, 1980d).[284]

---

281 See Section 7.1. "Biochemical Individuality and Its Implications".
282 See Section 5.4. *What to Do About Vitamins*.
283 See Chapters 7. The Scientific Study of Individuals and Alcoholism and 8. Genetotrophic Disease.
284 Williams later regrets his implication that members of *Alcoholics Anonymous* "often remain on the ragged edge and cannot trust themselves in the company of

9.3.1 Evidence

The 82 pages of *Nutrition and Alcoholism* contain fervent appeals to the medical community for the increased study of nutrition and alcoholism. Williams' greatest complaint in all publications regarding BI, generally speaking, is the lack of information on these various human characteristics and manifestations thereof. In other words, a lack of medical evidence on which a possible treatment could be based. Williams expresses thoughts akin to those nowadays summarised under the term "evidence-based medicine".[285]

> It would be unscientific in the extreme to claim that many diseases are curable by the nutritional means which we recommend. It is not unscientific, however, to make trials and find out. (…) The only path to progress in the medical field is experimentation and trial. (Williams, 1951a, p. 65)

This excerpt suggests the importance of basing medical treatment on reliable and trustworthy data from trials, rather than on the notions or experiences of single physicians. The proposal that "perfection of methods treating alcoholism will come as a result of research and experimentation" further accentuates this thought (Williams, 1951a, p. 78). Though EBM is a movement developed in the 1980s and 1990s at McMaster University in Canada (Howick and Glasziou, 2011, chap. 2), Williams expresses the need for increased appreciation of scientific research in medicine in 1951.

To Williams, a trained and experienced scientist by trade, the procedure of repeated trials to confirm or reject a given hypothesis is one immediately applicable to medical research. Williams himself describes his responsibility as a biochemist as the "discover[y of] the way our bodies work and to suggest new treatments for disease" in 1951 (Williams, 1951a, p. 7). In an "age of biochemistry" in medical research, the application of these discoveries is attributed great promise for improving medical therapy (Williams, 1951a, p. 7). This methodical

---

ordinary men, who in increasing numbers are using alcoholic beverages" (Williams, 1951a, p. 5).

285 "Evidence Based Medicine" is also referred to as "EBM" within this thesis. EBM can be defined as "the conscientious, explicit, and judicious use of current best evidence in making decisions about the care of individual patients. The practice of evidence based medicine means integrating individual clinical expertise with the available external clinical evidence from systematic research" (Sackett et al., 1996, p. 71). EBM "de-emphasizes intuition, unsystematic clinical experience, and pathophysiologic rationale as sufficient grounds for clinical decision making and stresses the examination of evidence from clinical research" (Guyatt et al., 1992, p. 2420).

scientific approach to medical research reflects the scientific zeitgeist of the early 1950s. With the first broadly published randomised controlled trials conducted in the previous decade, the Council on Pharmacy and Chemistry had suggested further such trials in medical research mere two years prior to the publication of *Nutrition and Alcoholism* (Council on Pharmacy and Chemistry, 1949).[286] With the concept of BI not yet established within the medical community, these excerpts press for increased appreciation of biochemical research by doctors. As Williams is convinced that the treatment of alcoholism (as is the case for all diseases) belongs in the hands of trained physicians (Williams, 1951a, p. 4), the appreciation of his theories by these medical professionals are essential to their application. He highlights his conviction of the essential soundness and efficacy of his approach (Williams, 1951a, p. 55).

The preliminary results of medical research on the nutritional treatment of alcoholism are presented within *Nutrition and Alcoholism*, providing the evidence which Williams calls for in all aspects of individuality (Williams, 1951a, chap. 8). Carried out by doctors in eight locations across the United States, the information presented in *Nutrition and Alcoholism* is based upon "irregular and informal" reports of physicians treating alcoholics through nutritional means (Williams, 1951a, p. 45). As these "will publish their own reports when the studies have been completed", Williams (1951a, p. 45) is only able to provide anecdotal information on those cases well-known to him. Of the approximately twenty patients regularly participating in the trials, Williams is able to present the stories of five alcoholics and the alleviation of their compulsion to drink through the nutritional treatment suggested within the book (Williams, 1951a,

---

286 The National Institute for Health and Care Excellence (NICE), the institute developing and publishing British medical guidelines, defines a randomised controlled trial as "a study in which a number of similar people are randomly assigned to 2 (or more) groups to test a specific drug, treatment or other intervention. One group (the experimental group) has the intervention being tested, the other (the comparison or control group) has an alternative intervention, a dummy intervention (placebo) or no intervention at all. The groups are followed up to see how effective the experimental intervention was. Outcomes are measured at specific times and any difference in response between the groups is assessed statistically. This method is also used to reduce bias" (National Institute for Health and Care Excellence, 2023a). The exact year in which the first study that could be defined as a randomised controlled study is contested. For further information on the history and development of medical trials, see Bull (1951), Collier (2009), Bhatt (2010), and Craft (1998). For further information on the work of NICE, see National Institute for Health and Care Excellence (2023b).

pp. 47–52). Of these twenty, all seem to benefit from treatment, even though some do not lose the compulsion to drink completely (Williams, 1951a, pp. 52– 53). Though this doubtlessly constitutes an achievement for the researchers at DBUT, Williams highlights the fact that these results are neither fully conclusive nor statistically unchallengeable (Williams, 1951a, pp. 54– 55). A further issue with the trials underway is that the physicians in question are often unsystematic in their approach to the research, limiting the suitability of the results (Williams, 1951a, p. 52). They do, however, fall in line with the results of animal experimentation, which are classed as "clear cut and unequivocal", therefore indicating a principle validity of the theory (Williams, 1951a, p. 54).

A final aspect regarding evidence in *Nutrition and Alcoholism* involves the example of BI which appears most often at the beginning of Williams' interest in this subject; the ability of the bloodhound to distinguish between individual human beings reappears in 1951.[287] Other than in previous instances, however, this phenomenon is *explained* rather than merely posed as an example of possible uniqueness. Describing biochemical research of the urine and saliva of individual human beings, Williams attributes this aptitude of the bloodhound to the proven chemical distinctiveness of bodily excretions (Williams, 1951a, p. 17). This research is, in fact, published later the same year as part of *Biochemical Institute Studies IV: Individual Metabolic Patterns and Human Disease: An Exploratory Study Utilizing Predominantly Paper Chromatographic Methods* (Kirby Berry and Cain, 1951). The confirmation of this notion, merely anecdotal evidence at the beginning of Williams' career in the research of BI, further underlines his conviction for the importance of evidence. Early experiences in which a lack of evidence forbade the logical explanations for certain phenomena remain a driving force even when the theory is much further developed.[288]

9.3.2 Supplementation

Within the construct of a genetotrophic origin of alcoholism, the subject of nutritional therapy options plays a central role. With the genetic aspects of such diseases difficult to control, the supplementation of necessary nutrients and vitamins offers the seemingly simplest and most accessible treatment. Though a plethora of diets and food plans furnishing rich sources of certain micronutrients and vitamins are widely known today, Williams' approach to the treatment

---

287 See Sections 5.3. Practical Applications of Vitamin Research and 6.5. "Individuals Vary Greatly".
288 See Section 4.1. Youthful Observations.

of such disease is not centred around the alteration of a patient's intake of food. While it should also be optimised, he indicates that a change in diet does not suffice to alleviate such paroxysms of alcohol consumption. As Williams is of the opinion that *"there is no food source rich enough to supply generous amounts of all* [B vitamins]", which are central in his approach to tackle alcoholism, the use of special supplement capsules becomes necessary (Williams, 1951a, p. 38). A capsule containing a very specific combination of 15 nutrients and vitamins, provided to Williams and his colleagues free of charge by Eli Lilly and Company, is recommended to meet all needs of most alcoholics (Williams, 1951a, pp. 39–40).[289]

The configuration of these tablets, designed by the researchers at DBUT, is based on previous experience with supposed human needs for the respective ingredients, special aspects of the metabolism of alcoholics, and practicability regarding the size of each individual pill (Williams, 1951a, p. 40). As has been demonstrated on multiple occasions, Williams' prior biochemical research is of essential help in the formulation of solutions to issues arising from BI. Though it is made clear that this in no way constitutes a failsafe answer to all issues of alcoholism, a concrete method of approach is put forward.

9.3.3 Appeal

An essential aspect of *Nutrition and Alcoholism* is the appeal for the acceptance and research of the Genetotrophic Concept. Williams, convinced of the validity and soundness of this theory, requires accepting and enthusiastic physicians to make the widespread investigation thereof possible. Though the research of rodents is easily possible within his laboratory, a proven validity of his suggested medical treatment can only arise from clinical trials conducted by medical doctors, not biochemists. Williams admits that the lack of support by physicians forces DBUT to conduct own clinical trials, for which the institute is not equipped (Williams, 1951a, p. 78). This missing acceptance by the medical community hampers the progress of Williams' research. Chapter 12 of the book from 1951 is indeed entitled "A Plea for Clinical Research", and impresses the importance of the study of individuals as well as pointing out three fundamental issues of clinical research (Williams, 1951a, chap. 12). These are the lack of prestige of those physicians dedicating their lives to research, the lack of economic security for the same physicians, and the discouragement of research by clinical

---

289   Eli Lilly and Company ("Lilly") is one of the world's largest pharmaceutical firms, founded in 1876. For further information on Lilly, see Eli Lilly and Company (2023).

teachers in medical schools (Williams, 1951a, p. 77). Williams repeatedly criticises medical education (and education in general) in the United States later in his career (Williams, 1957c, 1971, 1972, 1986). Through the publication of this book, and the great importance attributed to the study of the topics discussed within, he hopes to create public awareness for the relevance and pertinence for such research.

Having produced his own first evidence of the validity of these concepts, *Nutrition and Alcoholism* hopes to increase support for GC and the concept of BI. It is deemed probable that the result of "actual trials" will be "an increased conviction of the importance of individuality and the need for studying human beings after the manner suggested in *The Human Frontier*" (Williams, 1951a, p. 79). This appealing nature of *Nutrition and Alcoholism*, already relevant in *The Human Frontier* and restated in other medical and scientific journals (Williams, 1951c, 1951e),[290] is also taken up by its reviewers.[291] The publication of this book, therefore, comes at an essential time for Williams and his ideas, and is hoped to project the idea of BI into mainstream medicine and medical treatment.

9.3.4 Promotion and Reviews of *Nutrition and Alcoholism*

As is the case for the Genetotrophic Principle, the promotion of *Nutrition and Alcoholism* and the information contained within dominate the year following its publication.[292] Multiple articles and presentations document Williams' activities (The Editors, 1952; Williams, 1952a, 1952b). These are less numerous than those found for GC, due, in part, to the untimely death of Williams' first wife, Hazel, in 1952 (The New York Times, 1952). Following this event, Williams takes a leave of absence throughout the summer of 1952 and the first semester of 1953 (Rogers, 1952). The three published articles in this timeframe merely reiterate the abridged contents and point to *Nutrition and Alcoholism*.

---

290 See Section 6.1. "Humanity Must Understand Itself".
291 See Section 9.3.4. Promotion and Reviews of *Nutrition and Alcoholism*.
292 See Section 7.6. Symposium and Society.

June 9, 1952

Dr. T. J. Boman
Special Research Assistant
Public Health Laboratory
Poona - 1
Bombay, India

Dear Dr. Boman:

    Your letter of May 30th to Dr. Roger J. Williams has been received here at his office. You apparently have not heard of Mrs. Williams tragic death this past February. She had been in a very depressed state for sometime, and in February she took her life by her own hand. Dr. Williams is not on leave for the summer as well as for the first semester of next year.

    Our reprint supply on the research concerning alcoholism has long since been exhausted. Your best bet for getting the entire story would be to order Dr. Williams' book Nutrition and Alcoholism, University of Oklahoma Press, Norman, Oklahoma, price $2.00. This was published last year and you can order it directly from the publishers.

    The following is a list of journal references where this work has appeared:
1. Quart. Jour. Stud. Alcohol 7, 567-587 (1947)
2. Proc. Natl. Acad. Sci. 35, 265-271 (1949)
3. Arch. Biochem. 23, 275-290 (1949)
4. Texas Rpts. Biol. Med. 8, 238 (1950)
5. Arch. Biochem. 29, 27-40 (1950)

However, I doubt that you will have all of these available, and the book will probably be your best source for complete information.

                                           Sincerely yours,

                                           Lorene Lane Rogers, Ph.D.
                                           Research Scientist
                                           and Executive Assistant

LLR/jk

Fig. 14: Letter by Lorene Lane Rogers to T. J. Boman on 09.06.1952 indicating Williams' absence following Hazel's death; Dolph Briscoe Center for American History, The University of Texas at Austin, Roger Williams Papers, Box 88-087/23a, Folder: Alcoholism – Inquiries, Reports, Correspondence, Physicians, Scientists, Governmental Agencies, 1955 –

Nutrition and Alcoholism, reviewed in an array of newspapers and journals, is a largely well-received book. With *The Human Frontier* and multiple publications indicating his expertise within the field of individuality, no questions are raised as to the legitimacy of a biochemist's interest in such matters. A review in *JAMA* describes the Genetotrophic Concept as "provocative" and "intriguing", suggesting that the research of nutritional alleviation of disease offers an interesting avenue of research (Nutrition and Alcoholism, 1951). An aspect highlighted by this review, and taken favourably by its author, is that Williams "does not claim to present a universal cure for alcoholism" (Nutrition and Alcoholism, 1951). This facet of humility regarding ideas around BI differs starkly from his previous book *The Human Frontier*. In contrast to *The Human Frontier*, in which reviews criticise Williams' grand announcement of his solution to all issues of society, *Nutrition and Alcoholism* is much more modest regarding the ideas it presents.[293] Highlighting the lack of concrete evidence for the effectiveness of his suggested treatment, Williams pronounces that "supplying nutritional needs (…) is certainly not a means of eliminating all human problems" (Williams, 1951a, pp. 53–54).

A review in *The New York Times* by an instructor of pharmacology and therapeutics at Johns Hopkins University draws attention to the lack of support for controlled studies and addresses Williams' plea for "clinical and metabolic research on a high level" (Maren, 1951). It additionally takes up his request for societal change in making medical research attractive to doctors, with his approaches regarding individuality described as "sensible" (Maren, 1951). In the years following the publication of *Nutrition and Alcoholism*, DBUT receives thousands of letters from interested American and international parties, including alcoholics and their families, various companies, banks, doctors, academics, scientists, and private citizens, all having read of Williams' theories in journals and magazines.[294] The number of letters becomes so great, it necessitates the creation of standardised answers, as personalised responses would have become inordinately time consuming (Williams, 1958). Such reviews in a large medical journal and widely distributed daily newspaper as well as international journals (Fresneda, 1953; The Editors, 1953) exactly fulfil Williams' hope for increased attention to individuality in the medical field. They indicate an appreciation for

---

293 See Sections 6.1. "Humanity Must Understand Itself" and 6.6. Reviews of The Human Frontier.
294 These can be found in the Dolph Briscoe Center for American History at the University of Austin as part of the Roger Williams Papers in Box 88-087/23a.

the validity of Williams' research and disseminate his ideas and his concept for the nutritional treatment of alcoholism.

9.3.5 Summary

Nutrition and Alcoholism presents an abridgement of all information previously collected on nutrition and the aspects of alcoholism explored by Williams and his colleagues at DBUT. Alongside his appeal for the appreciation and creation of evidence regarding the individualities of human beings, Williams can present the first evidence of the validity of the nutritional approach to the treatment of alcoholism. The recommendation of a mixture of nutrients for the treatment of alcoholism, the supplementation of which provides the foundation for these first medical studies, is made based on Williams' previous experiences of vitamins and nutrition. The book functions as an appeal to the medical community and society at large to appreciate the issues of individuality and the Genetotrophic Concept. The positive reviews of *Nutrition and Alcoholism* indicate a growing interest in his concepts. Williams' studies on alcoholism continue in the following years, though they do not develop GC further and instead aim to optimise the treatment of alcoholics, attempting to prove the validity of his approach and confirm the validity of the Genetotrophic Principle (Rogers et al., 1955; Williams et al., 1955b).

9.4 Biochemical Institute Studies

With Williams' notoriety constituting an important factor in his efforts in promoting his own research, increased attention directed towards his ideas is further reflected in an invitation to speak before a medical conference of the *Muscular Dystrophy Associations of America* (Williams, 1953c). Williams here elucidates on the possibility of muscular dystrophy as a genetotrophic disease, presenting a diagram of polar coordinates depicting taste sensitivity, salivary constituents, and urinary constituents of twelve individuals (Williams, 1953c, pp. 119–120). These diagrams are chosen to indicate the extent of interindividual differences and reflect the research presented in DBUT's most extensive publication on Biochemical Individuality, *Biochemical Institute Studies IV* (Williams et al., 1951).[295] Appearing in May of 1951, this 207-page manuscript contains 21 articles, each presenting results from trials conducted at DBUT. Williams'

---

295 See Section 9.1. Biochemical Individuality V.

contribution is limited to the publication's opening article, summarising the results and preliminary conclusions of the research (Williams et al., 1951, Contents).

In his opening remarks, Roger Williams explains the thought process behind his interest in the biochemical study of alcoholism, rather than a broader approach involving multiple specialties. Though the interdisciplinary study of mankind continues to be his main objective, the biochemical groundwork study of interindividual differences is identified to be the most fruitful way forward following the publication of *The Human Frontier*. With interdepartmental study not yet possible, this biochemical research could help "pave the way for broader studies involving other disciplines" (Williams, 1951b, p. 8). With the focus of alcoholism identified, the realisation of the need to study "individuals in general" in order to make meaningful comparisons manifests itself. The concrete treatment options suggested in *Nutrition and Alcoholism* are described as unanticipated and unplanned at the outset of the investigation (Williams, 1951b, p. 8).

These general studies branch out into preliminary studies of schizophrenic patients, young children, individuals of different bodily composition (namely overweight and slender), children of reduced intelligence, and identical twins. The results of these "exploratory" studies (Williams, 1951b, p. 8), the preliminary and provisional nature of which is often highlighted, lead to six "tentative conclusions with 'far reaching [implications]'" (Williams, 1951b, p. 17), "not only into the field of medicine but also into the field of education and into the numerous ramifications of the problem of human behaviour" (Williams, 1951b, p. 18). These largely confirm the thoughts posed in prior publications on the topic of individuality. The distinctiveness of each individual from conception, the connection of metabolic patterns and susceptibility to disease, the necessity of a genetic understanding of disease, and the connection of metabolism and psyche are all theories posed previously, though indications for their truthfulness are only found through further research (Williams, 1951b, pp. 18–21). All other articles in *Biochemical Institute Studies IV* are published by researchers at DBUT without Williams' collaboration. As institute director, his role is limited to the introduction and resume of this publication.

An aspect of this work interesting from a sociological point of view is Williams' adamance that "*differences do not denote defectiveness*". Neither difference in anatomy, abilities, sex, or race are indicative of weakness or inferiority in any way, according to Williams. In fact, these differences are something to be cherished. They are the root of the human love for freedom, and the black and white thinking of "defective" and "normal" patterns is neither accurate nor fruitful (Williams, 1951b, p. 20). The issues of race or sex do not appear in prior publications and do therefore not play a role in BI up to this point. In 1951, more than a decade before the 1963 "March on Washington" of the Civil Rights Movement and three years before the famous Supreme Court Ruling on *Brown v. Board of Education* (Warren and Supreme Court of The United States, 1954),[296] [297] racial segregation and racist tendencies and views are still a norm in American society. In a pre-Civil Rights Movement era, Williams' views are doubtlessly progressive for a middle-aged, white, Texas-based, American male. The proclamation of a biological parity of all races is an indication for the ultra-societal nature of BI; differentiation of individuals according to unscientific societal ideals has no place in a concept seeking to overturn these. Though it is not possible to ascertain whether black patients are included in the studies of DBUT, it is notable that all differentiations according to ethnic groups (to which Native Americans and persons of, in Williams' terms, "Jewish" origin are also counted) in Williams' publications are of a purely statistical nature (Williams, 1947e, pp. 582–583, 1948c, p. 53, 1953a, p. 103). The appreciation of individuality is democratic in this sense: every individual is to be judged solely by his or her own individual measurements, abilities, wants, and needs. The issues of race and racism reappear in *Free and Unequal* and following publications.[298]

Furthermore, the appreciation of an equality of the sexes, with neither being described as superior to the other, does not reflect the societal norms of the early 1950s. With the ideal of domesticity propagated even within the advertisements and other cultural elements of the time, gender equality was not at the forefront of societal discussion.[299] Equally notable is, therefore, this propagation of biological parity between the sexes, with no aspect of talent, capabilities, psyche, or anatomy idealised or stigmatised. As fifteen of the twenty-two articles published in *Biochemical Institute Studies IV* are written either in collaboration with or in sole authorship of women, this parity of the sexes is also reflected within the publication (Williams et al., 1951, Contents). Though neither of these statements suffice to label Roger Williams as an early Civil Rights supporter nor an outright proponent of the feminist cause, it is noteworthy that the opinions propagated

---

296 For further information on "The March on Washington", see Jones (2013b).
297 For further information on *Brown v. Board of Education*, see Kluger (2004) and Klarman (2006).
298 See Sections 10.3.1. Communism and Racism Revisited and 11.4. Chemical Anthropology.
299 For critical appraisals of advertising and the role of women in the 1950s, see Gamber (2019), Catalano (2002), and Courtney and Lockeretz (1971).

by the research of individuality oppose the norms of the United States of the early 1950s.

9.5 Biochemical Individuality Following First Human Research

Many indications for the validity of the Genetotrophic Principle have been found through research at DBUT. Each human being has a metabolic personality unique to him or herself, and the metabolism of patients afflicted with alcoholism and other diseases differs markedly from healthy individuals. The trials of a nutritional supplementation scheme for the treatment, and possible prevention, of alcoholism are under way and show promising results.

9.6 Conclusion

For Williams, the early 1950s are characterised by the study of human beings, be it regarding alcoholism or their general metabolism. The formulation of a concrete treatment method for alcoholism based on prior experiences allows for the publication of *Nutrition and Alcoholism* and the study of the effectiveness of such treatment in real human subjects. Questions of evidence are raised, and fervent appeals toward the medical community made; the reviews of this book are positive in their tenor. In *Biochemical Institute Studies IV*, DBUT publishes a large quantity of research on individuality, including the development of new techniques for the large-scale study of human beings (Williams et al., 1951). Issues such as racial prejudice and gender inequality arise in first tentative discussions of the consequences of BI, with social betterment for all members of society its ultimate goal. Following such new insights of these studies and the death of his first wife, the next important publication further developing BI appears in 1953.

10 Free and Unequal

With studies producing data on human individualities underway at DBUT in the early 1950s, Williams once more turns his focus onto the non-scientific implications of individuality. The title of Williams' book *Free and Unequal: The Biological Basis of Individual Liberty* is suggestive of the political and societal issues discussed within (Williams, 1953a).[300] Published in 1953, it is yet another break in the grand scheme of his publications. While *The Human Frontier* attempts to tackle individuality from both a social and scientific point of view,[301] *Free and Unequal* places its primary focus on the importance of individuality within the humanities. Williams' previously voiced concern for the lack of research on individuality is reiterated, with its necessity illustrated in further detail. In this non-scientific book, Williams seeks to lay out the fundamentals of BI and its effect on American society in its various facets (Williams, 1953a, p. xii). The book seeks to "prove potent and stimulating if not revolutionary" to all those who read it, and builds upon the scientific evidence previously presented in *The Human Frontier* (Williams, 1953a, p. xii). Though unscientific in its approach, this publication corroborates the political facets of BI merely hinted at in previous articles and books. As a first entirely unscientific publication on BI, the substantiations of the previously more vague humanitarian consequences of Williams' theory must also be appreciated. The following pages discuss *Free and Unequal*, highlighting the new aspects of individuality it presents. As in the above, reviews are consulted to indicate the acceptance of the theories published in *Free and Unequal*.

10.1 Simple Yet Profound

Roger Williams' fourth major publication for the general public seeks to answer one question: "*Are newborn babies essentially uniform products?*" (Williams, 1953a, p. 3). The apparent simplicity of this query does nothing, however, to diminish the thoroughness in which Williams approaches his chosen topic. His prior warning that "simple questions may be profound and may hold complex implications" is fully justified by the elaborations of the following 171 pages (Williams, 1953a, p. 3). Drawing upon his own experiences in the study of

---

300 "*Free and Unequal: The Biological Basis of Individual Liberty*" is also referred to as "*Free and Unequal*" within this thesis.
301 See Chapter 6. *The Human Frontier*.

individuals, *Free and Unequal* discusses various aspects of human nature and composition, bolstering scientific evidence with personal accounts and anecdotes. The essential goal of this publication is to impress upon the reader the importance of heredity and its ramifications "as a practical and fundamental social factor" (Williams, 1953a, p. 6). This discussion is not lead on a purely academic or philosophical level, as the book is not solely aimed at scholars. The unscientific nature of *Free and Unequal*, noted by Williams (1953a, pp. xii, 6, 7) himself on multiple occasions, is further evidence by the fact that the book does not provide a single citation for the evidence of individuality it presents. This aspect is also noted in a later review (Anastasi, 1955, p. 243). In this case, the practical consequences and complications resulting from the inborn individuality of all human beings is of greater interest than those posed by theories and concepts.

The suicide of Williams' wife Hazel, which falls into the process of writing *Free and Unequal*, additionally influences the themes it discusses (Williams, 1953a, pp. 110–111). As he believes that the longevity and severity of her depression is prolonged and intensified by a lack of medical appreciation for her individuality, Williams sees the increased study of individuality as an opportunity to "prevent such calamities" (Williams, 1953a, p. 111). Describing the development of her illness in detail, Williams emphatically restates his previously voiced appeal to the medical community for the necessity of such study.

It becomes clear, therefore, that Williams' reasoning behind a wider discussion of the consequences of BI is varied. In gaining widespread acceptance of his ideas, Williams hopes to effectuate a societal "upheaval in basic thinking", and with it the improvement of the quality of life of all individuals (Williams, 1953a, p. 165). The rethinking, and consequent restructuring and rebuilding, of imperfect aspects of American civilisation could, in Williams' mind, further peace, increase the feeling of individual worth, and improve the freedom of all citizens (Williams, 1953a, pp. 11, 111). Following Williams' conviction that "[h]uman worth resides not only in those whom we regard as great, but in all of us, and we should provide an environment which will give everyone an *equal chance* to develop his potentialities in the way best suited to him individually", *Free and Unequal* is ambitious in its aims (Williams, 1953a, p. 12). The ongoing conflict of ideologies which characterise the Cold War additionally influence the themes discussed in *Free and Unequal*, revisiting subjects previously touched upon.[302]

---

302   See Section 10.3.1. Communism and Racism Revisited.

10.2 Signatures

Though the potential influence of BI on American society is the clear focus of *Free and Unequal*, its first chapters are dedicated to establishing a basic understanding of human individuality. Palpable examples of individuality and pronounced variations in human behavioural patterns are presented in order to depict the important role individuality plays in all societal interactions. As part of this evidence, Williams invites the reader to engage in a self-developed game which demonstrates the extensive variability of individuals' priorities (Williams, 1953a, chap. VI). Asked to rank 48 human desires according to his or her own preferences, the reader produces an entirely individual depiction of their own partialities and priorities in various aspects of life and society (Williams, 1953a, pp. 56–60). The results of this experiment carried out with ten subjects playing the game are graphed as polar coordinates and indicate the widespread differences existing within even modest cohorts. When comparing their own score with the average of the overall cohort, the magnitude of inter-individual variation is impressed upon the reader (Williams, 1953a, pp. 62–64). The overwhelming majority of examples of individuality presented in *Free and Unequal* also appear in prior publications and will therefore not be discussed further.

10.3 Politicisation

A number of Williams' prior works contain allusions to the political ramifications of BI, though the theme is rarely central.[303] Most previous publications have a clear focus on the scientific aspects of BI, practicing restraint in their comments on BI's humanitarian consequences. Conversely, *Free and Unequal* wishes to promote the application of Williams' theories to the betterment of American society, the scientific aspects of individuality merely presented as basic background knowledge. The political aspects are only hinted at and discussed briefly in previous publications. These, however, become a central theme in *Free and Unequal*, with large sections of the book dedicated to their analysis and critique. Williams' questioning of political ideology, race, and education bring an aspect of politicisation to BI previously not as prominent within the construct of the theory. *Free and Unequal* essentially demands the complete overhaul of large portions of American society to better furnish the understanding and appreciation of human individuality.

---

303 See Sections 6.3. Social Control, 7.5. Anti-Communist Sentiment, and 9.4. Biochemical Institute Studies.

10.3.1 Communism and Racism Revisited

Though hints of Williams' views on communism appear in previous publications, *Free and Unequal* unequivocally delineates the ways in which the principles of BI and this political ideology stand in opposition to one another.[304] The basic values of communist government, so Williams, are incompatible with a genuine appreciation for human individuality. These incompatibilities are discussed at length within the book, working titles of which include "The Hidden Roots of Communism" (Williams, 1951 f), "Of Genes and Freedom", "Freedom Has No Norm", "Freedom from Regimentation", and "The Biological Basis of Freedom" (Williams, 1952c). According to Williams, an end to the ideological clashes of the Cold War is only possible through the comprehension of Biochemical Individuality and a consequent revolution in Soviet society.

---

304 See Section 7.5. Anti-Communist Sentiment.

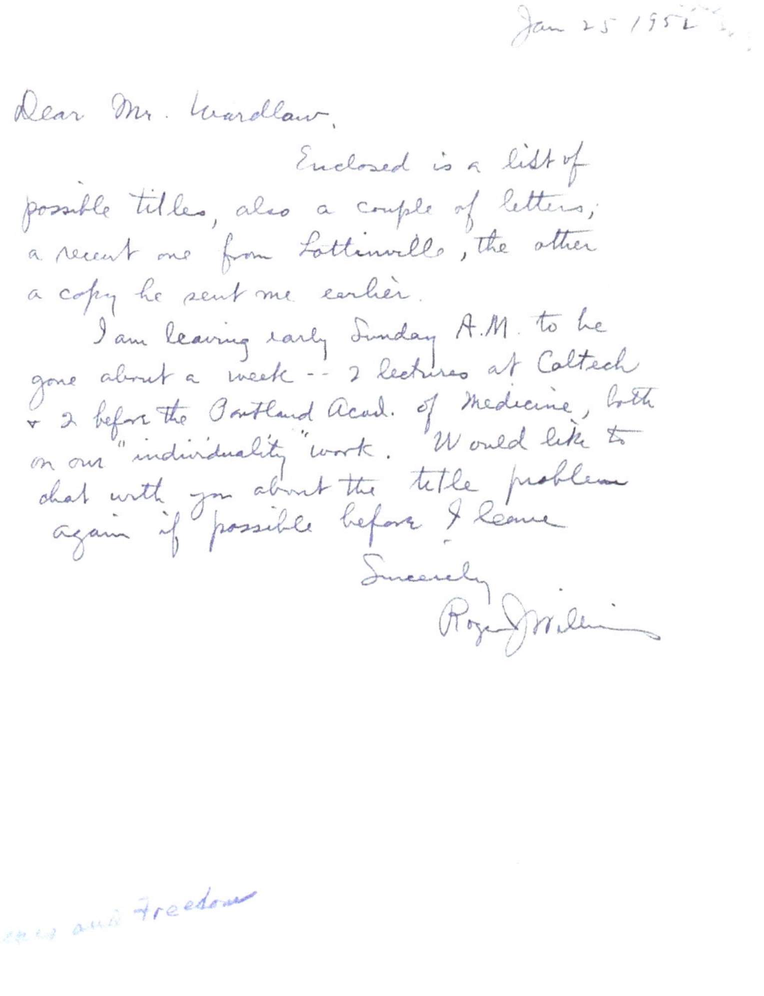

Jan 25 / 1952

Dear Mr. Wardlaw,

Enclosed is a list of possible titles, also a couple of letters; a recent one from Lottinville, the other a copy he sent me earlier.

I am leaving early Sunday A.M. to be gone about a week -- 2 lectures at Caltech + 2 before the Portland Acad. of Medicine, both on our "individuality" work. Would like to chat with you about the title problem again if possible before I leave.

Sincerely,
Roger Williams

Genes and Freedom

Fig. 15: Letter by Williams to Frank H. Wardlaw on 25.01.1952 including early ideas for the title of *Free and Unequal*, page one; Dolph Briscoe Center for American History, The University of Texas at Austin, Roger Williams Papers, Box 88-087/1k, Folder: The University of Oklahoma Press (Lottinville) 1950–1951

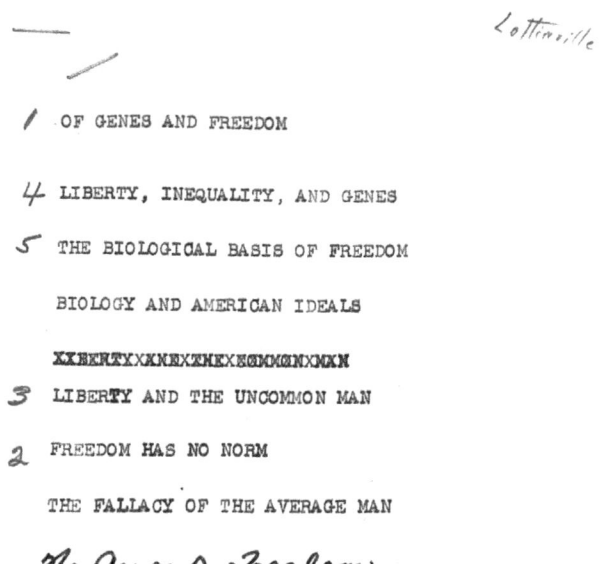

Fig. 16: Letter by Williams to Frank H. Wardlaw on 25.01.1952 including early ideas for the title of *Free and Unequal*, page two; Dolph Briscoe Center for American History, The University of Texas at Austin, Roger Williams Papers, Box 88-087/1k, Folder: The University of Oklahoma Press (Lottinville) 1950–1951

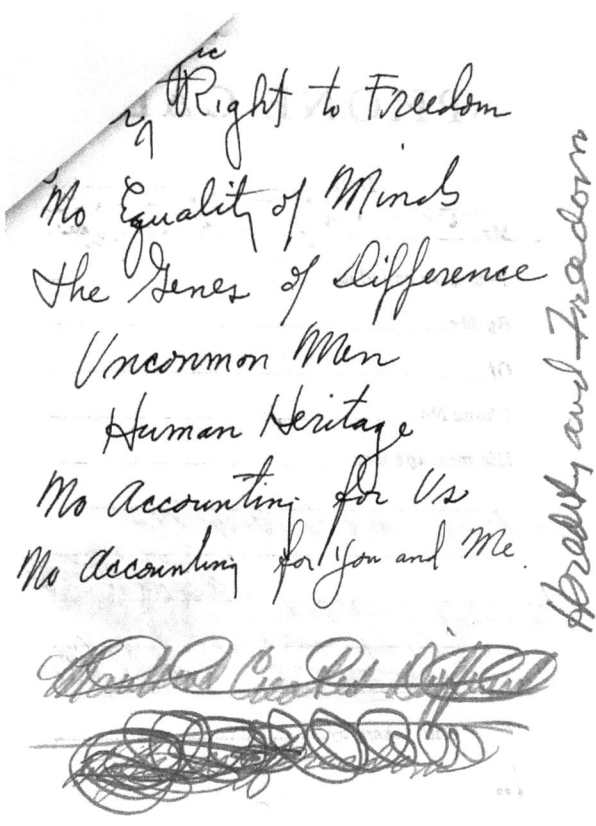

Fig. 17: Letter by Williams to Frank H. Wardlaw on 25.01.1952 including early ideas for the title of *Free and Unequal*, page three; Dolph Briscoe Center for American History, The University of Texas at Austin, Roger Williams Papers, Box 88-087/1k, Folder: The University of Oklahoma Press (Lottinville) 1950–1951

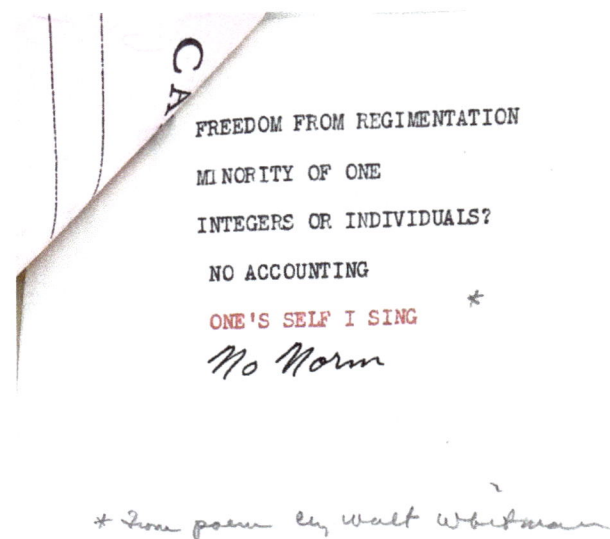

Fig. 18: Letter by Williams to Frank H. Wardlaw on 25.01.1952 including early ideas for the title of *Free and Unequal*, page four; Dolph Briscoe Center for American History, The University of Texas at Austin, Roger Williams Papers, Box 88-087/1k, Folder: The University of Oklahoma Press (Lottinville) 1950–1951

Discussing the liquidation of "free-thinking" scientists by "Soviet dictators", the contempt with which the ideals of communism are treated within this work immediately becomes evident (Williams, 1953a, p. xi). Soviet ignorance for the necessity of freedom, and its strong belief in a "uniformity doctrine" (Williams, 1953a, p. 142), are described as a "false premise" which fails to consider even the most basic of human needs (Williams, 1953a, p. 10). "The facts of human variability", conversely, are described to "constitute a firm foundation for our conceived ideal of democracy" (Williams, 1953a, p. 13). Indeed, Williams ascertains that "if the importance of human differences were to be stressed [in the USSR], a Pandora's box of troubles for their regime would be opened, leading to inevitable downfall" (Williams, 1953a, p. 10). Furthermore, Williams claims that an effective demonstration of the merits of appreciating BI is essential to ending the conflicts of the Cold War and abolishing pro-communist sentiment within the United States (Williams, 1953a, pp. 78–79, 144). With censorship of culturally American media an important policy in the Soviet Union, the publication of thousands of scientific studies of individuality in the international

Journal *Chemical Abstracts* is suggested as a potential method to "pierce the Iron Curtain" (Williams, 1953a, p. 160).[305] [306]

The issue of accepting a doctrine of uniformity is, however, not described as a purely communist flaw (Williams, 1953a, p. 142). As, so Williams, American children lack an appreciation for their individualities, an essentially communist ideal is deeply entrenched within even the society of the United States at the time, effectively hindering "liberty-loving democracy" to thrive (Williams, 1953a, p. 142). With the conclusion that "Americanism (...) is built upon the sand – a false idea of human nature", Williams wishes to "abandon it and build it upon a sound foundation" (Williams, 1953a, p. 11).

These examples indicate that Williams' criticism of the USSR and its communist ideals is much more pronounced in *Free and Unequal* than compared to previous publications.[307] Aside from a will to better the lives of American citizens, the suggestion of BI contributing to the efforts of winning the Cold War brings with it a new aspect of politicisation to the concept. With the characteristics of McCarthyism and societal anti-communism discussed previously still at play, BI is posed as a possible solution to the war of ideologies.[308] The first proxy-conflict of the Cold War is underway during the writing-process of *Free and Unequal*, the first draft of which is completed in the early months of 1952 (Williams, 1953a, p. 110).[309] With the destructive potential of the Cold War on display in Korea, Williams poses a possible non-violent solution to the conflict. The grand depiction of BI's potential importance further emphasises the enormous prospects Williams sees in his theory and his absolute conviction of their truthfulness, with

---

305 "From 1946 to 1989, "The Iron Curtain" was a nickname for the border between Western Europe and the communist countries of Eastern Europe, which made it very difficult to travel into or out of Eastern Europe (McIntosh, 2013). For further information on "The Iron Curtain", see Applebaum (2012).
306 *Chemical Abstracts* is the publication of the Chemical Abstract Service (CAS) of the American Chemical Society. It summarises and condenses thousands of scientific chemical journals and is published internationally. *Chemical Abstracts* was available in the USSR in 1953 (Williams, 1953a, p. 160). For further information on CAS, see Chemical Abstracts Service (2023).
307 See Section 7.5. Anti-Communist Sentiment.
308 See Section 7.5. Anti-Communist Sentiment.
309 The Korean War (1950–1953) is often described as the first war-by-proxy of the Cold War, with the United States and United Nations supporting South Korea (Republic of Korea) and the Soviet Union and The People's Republic of China supporting the North (Democratic People's Republic of Korea). For further information on the Korean War, see Hastings (Hastings, 2020) and Stueck (1995).

necessity for radical reform continually reappearing as a central theme throughout the book.

The second societal issue to be given greater attention in *Free and Unequal* is that of racism and the conflicts arising from racial inequality. While the incompatibility of BI and racist sentiment is first thematised in previous publications, these thoughts are developed to a further degree in *Free and Unequal*. Though the statements of previously discussed publications are merely suggestive, an entire chapter is dedicated to the "distressingly bad" state of racism in the United States in *Free and Unequal* (Williams, 1953a, chap. XIII, 1953a, p. 121). The main focus of this discussion lies on the plight of the African American community, Williams referring to this ethnic group as "Negroes" within the book.[310] Williams categorically and clearly positions his views on racial stereotypes and racist tendencies, rejecting the dogma of white supremacism.[311]

> No matter what group we consider, whether Caucasian, American Indian, Negro, or Mongolian, we will find that each member exhibits individuality; he will have his own "signatures", his own pattern of thinking, and his own set of wants and satisfactions. This makes the lumping together of all members of an ethnic group, as though they were the same, wholly unwarranted.
> (…)
> If there are racial *patterns*, these too cannot be called superior and inferior except with respect to single items. This leads us to the conclusion that there are no "superior" and no "inferior" races (Williams, 1953a, p. 122).

This excerpt confirms the notion of Williams' views on racism touched upon in *Biochemical Institute Studies IV*.[312] Going further, he suggests that African Americans are unquestionably "superior to whites in a number of ways", actively contradicting the doctrine of white supremacy (Williams, 1953a, p. 124). Within his concept of Biochemical Individuality, Williams wishes to promote the appreciation and celebration of all human individuality, regardless of race, gender, or

---

310   While the use of the term "Negro" or "Negroe" is outdated and considered racist today, it was a common term used to describe the descendants of former slaves and black people in the United States in the early 1950s. As such, it was considered a standard term for African Americans until the Black Power movement of the 1960s. It can, therefore, be assumed that Williams' use of the term is not intended to diminish or insult this ethnic group in *Free and Unequal*. For further information, see Stevenson (2010) and Martin (1991).
311   Movements promoting white supremacism purport the superiority of ethnically white individuals over people of colour. For further information, see Jenkins (2022).
312   See Section 9.4. Biochemical Institute Studies.

religion (Williams, 1953a, pp. 125–126). Such societal appreciation, so Williams, must not only acknowledge differences, but allow each race "equal opportunities for education, economic advance, and well-being" and encourage "*distinctive* contributions to culture" (Williams, 1953a, p. 128). The forced uniformity of any individual is, to Williams, a ridiculous concept, and the idea of imposing a specific mould on entire racial communities even more so (Williams, 1953a, p. 128). Williams recognises the difficulty of disbanding racial stereotypes, yet proposes that a study and understanding of all human variability could aid in the fight against racism (Williams, 1953a, p. 126).

With such statements, Williams clearly positions himself and the concept of BI as promoters of racial equality in all aspects. The understanding he wishes to furnish through the appreciation of the variations all humans show is, to him, incompatible with hatred on the grounds of ethnicity. These excerpts are further indications for the increased politicisation of BI developed in *Free and Unequal*, and exemplify the political message of acceptance and appreciation behind *Free and Unequal*, an aspect favourably appreciated in a later review of the book (Danforth, 1953, p. 404).

10.3.2 Assembly Line Educations, Regimentation, and Dogma

Within the construct of his societal appraisal, Roger Williams attaches great importance to the topic of education. It is here, so Williams, that the failures of society to appreciate individuality begin. Williams criticises the "universalized" scope of the American schooling system (Williams, 1953a, p. 71), which fails to adequately cater to the needs of individual children, essentially creating an "Educational Assembly Line" (Williams, 1953a, p. 68). "Education", so Williams, "(…) should expect to find each individual outstanding in some qualities that can be *educated*", rather than creating frustration in those not ideally suited to the chosen teaching methodology (Williams, 1953a, p. 73). The education of children in the facets of their own individuality is suggested as a solution to this problem (Williams, 1953a, p. 73). Throughout their school career, a student should "continuously learn more and more about himself and more and more about the society into which he has to fit (or misfit, if need be)" (Williams, 1953a, p. 73). This should, so Williams, "foster the love of freedom" characterising American ideals and therefore reduce the likelihood of children's infatuation with dictatorial structures (Williams, 1953a, p. 70).

Again, Williams suggests fundamental changes to an essential aspect of American society. The result of Williams' musings is the suggestion of a substantial overhaul of the American schooling system. The changes to the education of more than one-hundred and fifty million Americans that he demands are far-reaching and comprehensive (US Census Bureau, 1953, p. 9). This readiness to part with deeply entrenched and long-established systems within the United States based on his ideals of individuality further shows the importance he attributes to BI.

Williams' reflexions on ideological topics are relevant to the development of BI as they introduce an additional layer to his concept. While *The Human Frontier* suggests potential for societal improvement of social control and better international relations, BI is attributed an essential role in the abolition of all manner of societal conflicts in *The Human Frontier*. The critique of essential structures in the United States goes much further than the appeal for the increased study of individuals voiced in *The Human Frontier*. Williams even questions which laws can effectively govern a society of heterogenous individuals, suggesting that effectual law-making is only possible when the individuality of all human beings is taken into account (Williams, 1953a, p. 86). With individuality at the root of every human's will and need for freedom, Williams makes the case for restraint from over-regimentation in law-making, as it leads to a disregard for even the most sensible of laws. A further degree of freedom is suggested to be what individuals require (Williams, 1953a, p. 83).

The concept of uniformity is exclusively associated with negatively connotated aspects of society in *Free and Unequal*: communism, racism, and inadequate education. The repeated labelling of ideas involving human uniformity as "doctrine", to be believed in or rejected, creates an air of dogmatism, as do the repeated calls for fundamental societal reform (Williams, 1953a, p. 6). This doctrine, incorporated into the thoughts of "a great body of social psychologists, sociologists, social anthropologists, and many others, including historians, economists, educationalists, legal scholars, and men in public life", is seen as fundamentally flawed and worthy of a complete overhaul (Williams, 1953a, p. 8). The dogmatic nature of *Free and Unequal* is highlighted by Williams' assertion that "there is no middle ground" between his and other views on life and society (Williams, 1953a, p. 11). Having produced some, but not exhaustive, data on the individuality of humans, the evidence available to Williams is seen as sufficient to merit the demand of a complete reorganisation of education and extensive parts of American society.

As has been indicated by the previous sections of this chapter, Williams' central demand in *Free and Unequal* is a strict and resolute restructuring of the thought processes and political structures within the United States. The perceived arrogance shown by a refusal to accept other interests, opinions, and capabilities as adequate is indicated to be detrimental in all aspects of life and for everyone involved (Williams, 1953a, pp. 145–147). In this spirit, Williams admits that the study and appreciation of individuality is not "the only way to arrive at the truth" (Williams, 1953a, p. 153) and does not claim to provide infallible answers (Williams, 1953a, p. 171). Addressing his own life experiences, Williams feels confident that his satisfaction and happiness has increased through his understanding and acceptance of the individuals with which he interacts (Williams, 1953a, p. 171). Suggesting that his "revolution" of thought processes would require the rewriting of most publications ever written (Williams, 1953a, p. 168), the revolutionary essence of Williams' concept becomes more clear in *Free and Unequal* than ever before. This "upheaval in basic thinking", as he describes it, is much more far-reaching than the design of studies on human individuality and indicates the scale of necessary change (Williams, 1953a, p. 165). Picking up on material produced for an unpublished manuscript from 1950, William indicates how this rethinking must even take place even on a basic linguistic level (Williams, 1950d).[313]

10.4 Reviews of *Free and Unequal*

The previous sections have shown the radical nature of the changes which Williams suggests are necessary for American society and democracy to thrive in *Free and Unequal*. The book reviews appearing in a large variety of publications provide an indication for the acceptance of the thoughts and concepts posed by Roger Williams in 1953. They also suggest how wide a readership *Free and Unequal* reaches.

Having previously gained great distinction within the field of Biochemistry, the prominence of Williams' name in his field is interpreted as a mark of quality for *Free and Unequal*, with his expertise in the field of Biochemistry left unchallenged in all reviews available at the time of writing. Broad acceptance of the idea of non-uniformity is voiced by anthropologists (Montagu, 1953), geneticists (Danforth, 1953), education-specialists (Horn, 1954), psychologists (Paterson, 1954; Anastasi, 1955; Woodworth, 1955), and pharmacologists alike (Martin, 1956). All aforementioned reviewers agree on the fact that Williams' elucidation of individual differences for laymen is a most valuable contribution to science and society, though the notion that Williams slightly overstates his case is similarly ubiquitous (Anastasi, 1955; Danforth, 1953; Horn, 1954; Martin,

---

313 This manuscript can be found in the Roger Williams Papers, Box 88-087/41, Folder: Plea for temperance and discrimination in the generic use of the word 'the'.

1956; Montagu, 1953; Paterson, 1954; Woodworth, 1955). The work is described as "illuminating and stimulating" (Woodworth, 1955, p. 67), "interesting and provocative" (Free and Unequal, 1953), "eminently humane and [making] his principal points powerfully" (Montagu, 1953), and "presented more effectively (…) than psychologists themselves are able or willing to do" (Paterson, 1954, p. 642). With such positive reviews published in a wide range of public newspapers (Montagu, 1953), scientific journals (Anastasi, 1955; Danforth, 1953; Free and Unequal, 1953; Martin, 1956; Paterson, 1954; Woodworth, 1955), academic publications (Horn, 1954), and teachers' journals (Ballou, 1955), the widespread esteem in which *Free and Unequal* is held becomes evident. The reissue of the book in 1964 and 1979 similarly indicates an appreciation of its theories and contents beyond its own time (Williams, 1964, 1979).

One prominent review of *Free and Unequal* is, however, pronounced in its criticism of Williams' grandiose depiction of his concepts of individuality. In an otherwise homogenous collection of overwhelmingly positive appraisals, psychologist Anne Anastasi (1908–2001) takes issue with Williams' depiction of his own ideas as new and even revolutionary (Anastasi, 1955, p. 244).[314] Though commending the book for its "vivid and popularized [depiction of interindividual differences] for lay readers", the information it portrays is described to merely comprise the elementary knowledge of a beginners psychology course (Anastasi, 1955, p. 244). Additionally, according to Anastasi, the aspects of environmental influence are given too little attention (Anastasi, 1955, p. 245). It is concluded, "that the positive contributions of this book are so inextricably intertwined with fallacious and unfounded statements as to render its net value questionable" (Anastasi, 1955, p. 246). Though this review is harsh in aspects of its critique, its negative view remains the exception.

10.5 Biochemical Individuality Following *Free and Unequal*

Biochemical Individuality is not only of consequence to scientific and medical research but is also central and essential to all personal and societal interactions. Society does not cater to the needs or wishes of individuals and must, therefore, undergo strict and thorough reform. Issues such as racism, failures in education,

---

314 Anne Anastasi was a famous psychologist and is most renowned for her psychological texts on testing in psychology. Recipient of the National Medal of Science and American Psychological Association President, she is still celebrated within her field today. For further information, see O'Connell and Russo (1988) and Pickren (2009).

law-making, and conflicts of ideology can be exacerbated by the doctrine of uniformity, which is entirely inaccurate.

10.6 Conclusion

The answer to the question posed by *Free and Unequal* is predictable to those aware of Williams' prior publications: babies differ from one another in the same way all adult human beings are unique. A lack of appreciation for this individuality is posed as the source of much strife and inequality in society, as well as being the cause for the ideological conflict of the Cold War and racial disputes. Presenting evidence for human individuality in numerous facets, *Free and Unequal* largely concentrates on the structural and political deficiencies arising from an acceptance of the idea of human uniformity. BI is attributed dogmatic qualities through the vehement assertion of its central importance and the rejection of all theories with contrary views. In this book, the far-reaching political aspects of BI are elucidated, and the case is made for the necessity of societal change.

11 Practical Genetotrophism

The final three years before the publication of *Biochemical Individuality* are characterised by an increased focus on the practical applications of the Genetotrophic Principle. With the theoretical aspects of GC discussed at length in years prior, the real-world ramifications and potential the theory holds come forth. This chapter discusses the new aspects introduced by Williams and his colleagues regarding GC following the publication of *Free and Unequal* in 1953, up to and including 1955. As *Biochemical Individuality* is published in 1956, his publications from this year and onwards will not be discussed. Because a first manuscript of *Biochemical Individuality* is completed by the end of 1954 (Williams, 1954c), the publications discussed in this chapter are doubtlessly influenced by Williams' research for his later book. They are, however, treated separately, as they contain indications for further developments to BI.

11.1 Genetotrophic Supplementation

The importance of adequate nutritional supplementation is a theme which reoccurs throughout the career of Roger Williams. With the effect of diets deficient in certain vitamins and nutrients at the root of his original biochemical studies,[315] adequate nutrition, the approximate requirements of human beings, and the formulation of a standardised supplement catering to these human needs are central themes of his work.[316] Having previously suggested a supplement for the nutritional therapy of alcoholism in *Nutrition and Alcoholism*,[317] Williams and his colleague Lorene L. Rogers (1914–2009) propose a supplement for the universal prophylaxis of "diseases of obscure origin" in 1953 (Williams and Rogers, 1953, p. 576).[318] The indications for such supplementation are suggested to be threefold: the compensation of poor nutrition, supplementation in conditions with increased requirements (pregnancy or previous diseases),

---

315 See Chapter 5. Vitamin Studies.
316 See Sections 5.3. Practical Applications of Vitamin Research, 5.4. *What to Do About Vitamins*, and 9.3.2. Supplementation.
317 See Section 9.3.2. Supplementation.
318 Lorene L. Rogers was a chemist and later served as the first female president of The University of Texas at Austin. For further information, see Hevesi (2009) and Office of the President at The University of Texas at Austin (2023).

and the furnishing of individual requirements due to metabolic individualities (Williams and Rogers, 1953, p. 574).

Though suggesting a personalised approach to be ideal, the formulation of a more generalised policy is necessitated by the lack of scientific methods able to measure individual nutritional needs. A chart including the estimated average daily need for 47 vitamins and minerals is presented, with dosages offered for those elements deemed worthy of supplementation (Williams and Rogers, 1953, p. 576). These values differ significantly from those suggested for the treatment of alcoholism in *Nutrition and Alcoholism* (Williams, 1951a, p. 39), as well as deviating from the suggested average intake of vitamins in *What To Do About Vitamins* (Williams, 1945a, p. 30). This indicates Williams' progress in research between the two publications. These suggestions, though advocating a practical approach for the prevention of genetotrophic disease, are merely theoretical in their nature, as such a supplement is neither advertised to exist, nor is any experimentational data legitimising its use offered. A more concrete suggestion for supplementation is posed in the case of muscular dystrophy, in which an increased intake of specific amino acids and vitamins is advised (Hurley and Williams, 1955, p. 390). Building on biochemical knowledge of vitamins and nutrition, these values are offered to promote good health and prevent genetotrophic disease.

11.2 Cancer

Though an interest in the fundamental biochemical research of cancer is originally at the root of DBUT's formation (Williams et al., 1966, p. 2), the study of tumorous tissues only rarely appears in Williams' publications. His lack of participation in such studies at DBUT is indicated by *Biochemical Institute Studies V*, which is entirely dedicated to the research of cancer, yet does not include any collaboration or direct association with Williams (Taylor et al., 1953). Though involved in its precursor publication, *Cancer Studies*, in 1945, Williams focuses entirely on the research and promotion of genetotrophic disease eight years later (Williams et al., 1945). While still partaking in the discussion regarding the research of cancer by reviewing another publication on the topic, he merely remarks upon the lack of appreciation for individuality within the reviewed book (Williams, 1953d, p. 234). This demonstrates that the extent of Williams' exclusive focus on BI in this period of his career goes as far as the exclusion of topics previously highly intriguing to him.

11.3 Individual Anatomies and Compositions

With the practical aspects of BI at the forefront of Williams' focus between 1953 and 1956, the need for more simple and illustrative examples of individuality arises. Though individualities in anatomy are sporadically and vaguely referenced in previous works, these differences are given much more thought from 1954 onwards.[319] In a later publication, Williams claims that the understanding of the dimensions of individuality shown by humans regarding their inner anatomy proved an essential insight for his understanding of BI (Williams, 1977, p. 57). The information in question is rooted in an *Atlas of Human Anatomy*, published in 1950 (Anson, 1950). This atlas not only portrays the most common anatomical variant of miscellaneous organs and bodily systems, but provides information on several variations for each singular compartment as the first of its kind. In citing a private conversation with its author, anatomist Barry J. Anson (1894–1974),[320] Williams shows that anatomical variants previously seen to be the "standard" are merely found to apply to 15 % of human specimens studied (Williams, 1954e, p. 332). The significance of this publication is also reflected by the multiple images from said atlas reprinted in *Biochemical Individuality* (Williams, 1956a, pp. 21, 22, 24, 25, 26, 27, 29, 31, 38, 42). An entire chapter is dedicated solely to anatomical variation in *Biochemical Individuality*, further indicating its significance (Williams, 1956a, chap. IV). While other studies of anatomical variants are cited in *Biochemical Individuality*, none is as present as Barry Anson's publication, which Williams discusses for the first time at a symposium in 1954 (Williams, 1954e). The description of anatomical individuality is embedded within the wider discussion of BI's relevance to law and judicial matters, though it offers no new aspects outside of those discussed previously elsewhere. Here, he summarises his view on individuality in constitution, which is also reflected in the contents of *Biochemical Individuality*'s fourth chapter:

> The picture of a man who has an average stomach, an average heart, average nerve patterns, average branching of the aorta, average sex glands, average thyroids, average pituitaries, average adrenal cortices, average amount of islet tissue in the pancreas, average taste reactions to various substances, average peripheral vision, average sensitivity to

---

319 See Sections 6.4. "Distinctive Metabolic Traits" and 9.4. Biochemical Institute Studies.
320 Barry Joseph Anson was an anatomist at Northwestern University and the University of Iowa. He is best known for his publication *An Atlas of Human Anatomy*, in which the manifold anatomical variation shown by individual human beings is depicted (Anson, 1950). For further information, see Arey (1962) and McCabe (1975).

pain in all areas of the body, average enzyme activities of various kinds, average rote memory, average arithmetical facility, average visual imagery, average liking for reading, music, sex, alcoholic beverages, conversation, etc., is purely a figment of the imagination. No such individual exists; no one remotely approaching this description exists; and if he did, he would be a most unusual individual – a most "abnormal" one. (Williams, 1954e, p. 341)

It is additionally noteworthy that this symposium-contribution is the first time Williams uses the term "humanics" in a publication for seven years (Williams, 1954e, p. 328),[321] also constituting the final usage of the term up to the publication of *Biochemical Individuality*.

Aside from such macroscopic anatomical differences, Karl Lashley's (1890–1958) research on the variations shown in human neural structures and their relation to human behaviour becomes a relevant theme in the appeal for the appreciation of BI (Williams, 1954e, p. 333, 1954g, p. 31).[322] Human beings begin their lives with entirely unique brain structures, showing microscopic as well as macroscopic variations (Williams, 1954g, p. 31), and exhibit additional definite and reproducible differences in the biochemistry and enzyme functions of these configurations (Williams, 1954g, p. 32). This not only begs the question of individual susceptibility to neurological and psychological illness, but of the entire medical concept of personality and its disorders (Williams, 1954g, p. 32). With nutritional means suggested as a potential treatment option should the underlying disease be caused by nutritional defects, Williams offers the solution he proposes for all little understood ailments: adequate supplementation. With individuality in anatomical features duly appreciated, Williams continues to develop further aspects of BI and GC in following publications.

Though the appreciation of macroscopic individualities is facilitated by the human capacity for sight, the microscopic composition of individuals may vary by factors of more than 50 without obvious conspicuities (Williams, 1954 f, p. 200). In 1954, Williams presents his view on the potential of a genetotrophic approach in the treatment of alcoholism at a symposium on resistance to toxic agents (Williams, 1954 f). Though he does not present new aspects of the Genetotrophic Principle, new evidence for the individuality of human beings on a microscopic level is brought forward. Collecting data from a total of 22 papers, Williams presents evidence showing wide variations in enzyme, amino

---

321 See Section 7.1. "Biochemical Individuality and Its Implications".
322 Karl Spencer Lashley was a well-known experimental psychologist, who dedicated much of his career to the study of mammalian behaviour in relation to brain functions (Carmichael, 1959, p. 129). For further information, see Carmichael (1959).

acid, and lipid concentration of human blood samples and the weight of the pituitary and other endocrine glands (Williams, 1954 f, pp. 197, 200, 201).[323] These differences in enzyme concentration range from variations of two- to fifty two-fold and in two cases an infinite value of variation between the minimum and maximum values of the cohort, while the mean factor of variation stands at twelve (Williams, 1954 f, p. 200).

Other tables show similarly large ranges in concentration and gland weights. Also used as evidence in various chapters of *Biochemical Individuality*, these values provide a further factual basis for the concept of BI. They indicate a profound individuality at the smallest level, not related to extraneous factors, but rather to inherent genetic differences. Additionally, Williams picks up on prior research of his own, discussing the results obtained from his study of alcoholics. Having found variations in the blood, urine, and salivary constitutions of individuals four years previously (Williams et al., 1950a) – the resulting polar coordinates of which are also reprinted in 1954 (Williams, 1954 f, pp. 198–199) – Williams now uses this evidence to discuss differences in constitution.

The existence of variation on a macroscopic and microscopic level are discussed and in part even proven to exist by Williams in prior publications.[324] In contrast, the collection of tables in "The Genetotrophic Approach to Alcoholism" is the first to underline the previously voiced deductions with extensive factual evidence from multiple external scientists and institutions. The use of external evidence is a central aspect of *Biochemical Individuality*, where Williams largely uses colleagues' published results to bolster his claims, rather than relying on only the results of his own research to convince his readers. While Williams' critique of the lack of evidence regarding individuality has been discussed in previous sections of this dissertation,[325] the increased use of external evidence for

---

323 Though Williams quotes all other referenced scientific data in this and other articles, the tables which present the data discussed above are entirely without reference. The exact values found in these tables, however, reappear in different chapters of *Biochemical Individuality: The Basis for the Genetotrophic Concept*. The values from Tables 1 and 2 can be found in Chapter IV of *Biochemical Individuality*, while Table 3 is reflected in Chapter V. The values in Table 5 can be found in Chapter VI. The full list of overlapping references can be found in Tab. 2: Articles Cited by *Biochemical Individuality* Appearing in Prior Publications.
324 See Sections 9.4. Biochemical Institute Studies.
325 See Sections 5.2. The Vitamin Content of Tissues, 5.4. *What to Do About Vitamins*, 6.1. "Humanity Must Understand Itself", 6.6. Reviews of The Human Frontier, 7.4. An Introduction to Biochemistry, Second Edition, 9.3.1. Evidence, 9.3.4. Promotion

human individuality indicates the growing body of information available on this subject.[326] Williams previously highlights the pressing need for further research, and the apparent increase in the amount of information available on the subject allows him to focus on less specialised papers and books encompassing a larger realm of topics, as he is no longer alone in studying individuality on a scientific level. The extensive citation and use of images from *Atlas of Human Anatomy* also indicates this shift; the exhaustive research done by other scientists allows for more effective and evidence-based publication.

11.4 Chemical Anthropology

Having coined multiple scientific terms during his career and devised a number of novel research methods, Williams introduces the scientific community to yet another new conceptual line of scholarship in December of 1954. Presenting his findings before the California Section of the American Chemical Society, of which he becomes president in 1957 (Davis et al., 2008), Williams discusses a novel term publicly for the first time (Williams, 1955a). Having conceived the concept of "chemical anthropology" two months previously, his new scientific line of research is to base itself on social, cultural, and physical anthropology (Williams, 1955a, p. 68). Discussing the importance of interindividual variation with relation hereto, this new brand of scientific research is largely akin to the concept of "humanics" discussed in previous chapters.[327] Indicating the possibility of revealing the "basis for harmonious human relations", Williams promotes the importance of chemical research to the increased understanding of human beings (Williams, 1955a, p. 80). Press cuttings announcing Williams' successful bid for presidency of the American Chemical Society, the largest scientific group in the world at the time (Fowler, 1988), largely allude to his discussion of "chemical anthropology", indicating the importance attributed to this speech and the potential described therein (Dr. Williams, '57, President of ACS, 1955; Texas Professor Heads American Chemical Group, 1955; Texas Professor Honored as Head Of Chemist Group, 1955; University of Texas Prof is Named New President of ACS, 1955; US Chemists Name UT Prof President, 1955; UT Biochemist To

---

and Reviews of *Nutrition and Alcoholism*, 10.1. Simple Yet Profound, and 10.6. Conclusion.

326 The focus here lies on human individuality. Research of the individuality shown by animals carried out by other scientists has previously been cited by Williams at this point.

327 See Chapter 6. *The Human Frontier*.

Head Society, 1955; UT Professor Elected Head American Chemical Society, 1955). Though this speech indicates a great interest in such studies, chemical anthropology does not appear in any additional research or discussions within the scope of this thesis, including *Biochemical Individuality*. Though it will therefore not be discussed further here, it does, however, indicate Williams' fervour to an increase of the studies of human individuals.

Aside from the subject of this specific talk, Williams' activities as a public speaker additionally indicate his established notoriety within the scientific community by 1955. Alleging he was merely voted into the presidency of the American Chemical Society due to his well-known publications, the importance of promoting his work and research is not lost on Williams (Hodge and Williams, 1980b). He speaks before a large number of societies and organisations in 1955, presenting his talk on chemical anthropology before 14 different Sections of the American Chemical Society in October of 1955 alone (Benner, 1955). This further increases his prominence and promotes the work most important to him: individuality.

11.5 Normal Young Men

The existence of multiple studies which visually elucidate the individualities of human beings through polar coordinates has been discussed in prior sections.[328] These polar coordinates emphasise subtle differences in metabolism by visualising even minimal variations which may be ignored or missed when presented in tabular form. In 1955, Williams publishes an article containing further such polar coordinates entitled "Metabolic Peculiarities in Normal Young Men as Revealed by Repeated Blood Analyses" (Williams et al., 1955a). This paper elucidates the results of detailed and repeated blood analyses performed on twelve young men without any known health issues. In line with previous publications, Williams presents his case for the Genetotrophic Concept and the importance of the study of individuals. Furthermore, information on the nineteen factors studied in all individuals is offered with the results presented in the aforementioned polar coordinate graphs (Williams et al., 1955a, pp. 617–618). As in the previous instances in which polar coordinates are presented, the striking variation between these individuals becomes clear.

This publication is distinctive because it presents the first true piece of research on healthy individuals which Williams repeatedly highlights as essential to the

---

328 See Sections 8.1. The Metabolic Individualities of Rats, 9.4. Biochemical Institute Studies, and 10.2. Signatures.

future of science and society. Though repeated analyses of patients with various ailments, such as alcoholism and muscular dystrophy, have occurred previously, this research has always focussed on finding a biochemical correlation between disease and variations in metabolism.[329] This research on "normal" young men, however, offers more conclusive proof of the metabolic peculiarities shown even by healthy individuals. Published in *Proceedings of the National Academy of Sciences of the United States of America*, this paper is aimed at scientists of all fields. It presents human individuality as a promising and attractive field of scientific study.

11.6 The Concept of Biochemical Individuality Before *Biochemical Individuality*

Every human being is an entirely unique individual. Be it macroscopic differences in anatomy, or microscopic and biochemical variances, every single aspect of an individual differs from his or her peers. These differences have been proven to exist in studies carried out by scientists of multiple fields and may be the basis of many diseases of obscure origin. Many of these diseases may be genetotrophic in nature and could, therefore, be successfully treated by nutritional supplementation. Applying the knowledge gained through the study of individuals can be the basis of solving countless issues facing society.

11.7 Conclusion

With the theoretical existence of a Biochemical Individuality conclusively proven to exist in mammals and humans, the final years prior to the publication of *Biochemical Individuality* are characterised by a search for further concrete evidence and practical examples of BI. With the practical application of the obtained knowledge as his focus, Williams formulates a possible nutritional supplement for diseases of obscure origin. Furthermore, the research of other scientists regarding BI becomes highly relevant in Williams' publications, with focus lying on macroscopic anatomical and microscopic differences in enzyme biochemistry. "Chemical anthropology" denotes a further diversification of Williams' research interests, alongside his increased notoriety as President of the American Chemical Society in 1957. Finally, Williams presents his own research on differences between healthy young males.

---

329   See Sections 9.3. *Nutrition and Alcoholism* and 9.4. Biochemical Institute Studies.

12 Biochemical Individuality

Biochemical Individuality: The Basis for the Genetotrophic Concept is published in 1956, 25 years after Roger Williams' first publication containing allusions to the concept of Biochemical Individuality was circulated.[330] Within this timespan, his research focus has been shown to transform from fundamental biochemical studies of vitamins to the science of individuals and their varied dissimilarities. The previous chapters have elucidated the alterations to and development of said theory, which finds its culmination in the publication of *Biochemical Individuality*. All previously discussed aspects of BI reappear within these 209 pages, bolstered by additional examples and discussions. The central message of this book is cogent and clear: human beings show a wide range of individualities and recognising these is of the absolute essence for science, medicine, and society. As with multiple previously published books, *Biochemical Individuality* is intended for the general public.[331] Once more aiming to make the concept of BI more approachable to all members of society, it contains illustrations of the most rudimentary biological principles on which it is based. As the culmination of 25 years of research, the book largely restates what has been discussed in previous chapters of this dissertation. Considerations in this chapter are therefore limited to new or altered aspects of BI. Additionally, articles published by Williams in 1956 are not considered separately. Their contents are either unrelated to BI or essentially identical with corresponding passages in *Biochemical Individuality* (including bibliographies), and they are thus not treated separately to the corresponding sections of the book itself.[332] Additionally, the preface of *Biochemical Individuality* dates it to March 1st of 1956, the vast majority of articles from 1956 are published after that date (Williams, 1956a, p. xi).

---

330 See Section 4.5. "'Taste Deficiency' for Creatine".
331 See Sections 5.4. *What to Do About Vitamins* and 9.3. *Nutrition and Alcoholism*, as well as Chapters 6. *The Human Frontier* and 10. *Free and Unequal*.
332 See Tab. 3: Overlapping References from *Biochemical Individuality* and Publications in 1956.

Tab. 3: Overlapping References from *Biochemical Individuality* and Publications in 1956

No.	Citation in *Biochemical Individuality*	Page(s) of Citation in *Biochemical Individuality*	Other Publications Containing the Identical Citation	Page(s) of Citation in Other Publications
1	Ralph S. Banay, Quart. Studies Ale., 4, 580–605, 1944.	108	Biochemical Genetics and its Human Implications (1956)	175
2	R. W. Engel, Proc. Soc. Exptl. Biol. Med., 52, 281–282, 1943.	156	Human Nutrition and Individual Variability (1956)	19
3	R. F. Light and L. J. Cracas, Science, 87, 90, 1938.	149	Human Nutrition and Individual Variability (1956)	17
4	Albert F. Blakeslee, Science, 81, 504–507, 1935.	127	Biochemical Genetics and its Human Implications (1956)	171
5	Albert F. Blakeslee and Theodora Nussman Salmon, Proc. Natl. Acad. Sci. US., 21, 84–90, 1935.	127	Biochemical Genetics and its Human Implications (1956)	171
6	Arthur Grollman, Essentials of Endocrinology, J. B. Lippincott Co., Philadelphia, Pa., 2nd ed., 1947.	81	Biochemical Genetics and its Human Implications (1956)	164
7	Max A. Goldzieher, The Endocrine Glands, D. Appleton-Century Co., New York, N. Y. and London, England, 1939.	90	Biochemical Genetics and its Human Implications (1956)	165
8	W. W. Jetter, Am. J. Med. Sci., 196, 475 (1938).	108	Biochemical Genetics and its Human Implications (1956)	172
9	Curt P. Richter, Quart. J. Studies Alc., 1, 650–662, 1941.	109, 218	Biochemical Genetics and its Human Implications (1956)	171

Tab. 3: Continued

No.	Citation in *Biochemical Individuality*	Page(s) of Citation in *Biochemical Individuality*	Other Publications Containing the Identical Citation	Page(s) of Citation in Other Publications
10	John M. Nagle, J. Allergy, 10, 179–181, 1939.	108	Biochemical Genetics and its Human Implications (1956)	172
11	Barry J. Anson, Atlas of Human Anatomy, W.V. Saunders Co., Philadelphia, Pa. and London, England, 1950.	22 and throughout Chapter 3	Biochemical Genetics and its Human Implications (1956)	163
12	K. S. Lashley, Psychological Reviews, 54, 333–334, 1947.	44	Biochemical Genetics and its Human Implications (1956)	165
13	Icie G. Macy, Nutrition and Chemical Growth in Childhood, Charles C. Thomas, Springfield, Ill., and Baltimore, Md., Vol. I, 1942.	103, 137	Human Nutrition and Individual Variability (1956)	12
14	E. Rissel and F. Wewalka, Klin. Wochschr., 30, 1065–1069, 1952.	61	Human Nutrition and Individual Variability (1956)	20
15	E. Rissel and F. Wewalka, Klin. Wochschr., 30, 1069–1073, 1952.	61	Human Nutrition and Individual Variability (1956)	20
16	Leland C. Clark, Jr., and Elizabeth Beck, J. Pedia., 36, 335–341, 1950.	69	Biochemical Genetics and its Human Implications (1956)	167
17	G. E. Hall and C. C. Lucas, J. Pharmacol, and Exptl. Therap., 61, 10–20, 1937.	71	Biochemical Genetics and its Human Implications (1956)	167
18	Michael Somogyi, Arch. Internal Med., 67, 665–679, 1941.	72	Biochemical Genetics and its Human Implications (1956)	167

(*continued*)

Tab. 3: Continued

No.	Citation in *Biochemical Individuality*	Page(s) of Citation in *Biochemical Individuality*	Other Publications Containing the Identical Citation	Page(s) of Citation in Other Publications
19	Gregory Pincus and Kenneth V. Thimann, eds. The Hormones, Academic Press, Inc., New York, N. Y., 1948, Vol. I.	85	Biochemical Genetics and its Human Implications (1956)	165
20	G. C. Ring, et al., J. Applied Physiol., 5, 99–110, 1952.	28	Biochemical Genetics and its Human Implications (1956)	164
21	William H. Rustad, J. Clin. Endocrinol. Metabolism, 14, 87–96, 1954.	43	Biochemical Genetics and its Human Implications (1956)	165
22	Robert G. Tucker and Ancel Keys, J. Clin. Invest., 30, 869–873, 1951.	53	Biochemical Genetics and its Human Implications (1956)	166
			Human Nutrition and Individual Variability (1956)	21
23	David M. Kydd, Evelyn B. Man, and John P. Peters, J. Clin. Invest., 29, 1033–1040, 1950.	53	Biochemical Genetics and its Human Implications (1956)	166
			Human Nutrition and Individual Variability (1956)	21
24	S. B. Barker, and M. J. Humphrey, J. Clin. Endocrinol., 10, 1136–1141, 1950.	53	Biochemical Genetics and its Human Implications (1956)	166
			Human Nutrition and Individual Variability (1956)	21

Tab. 3: Continued

No.	Citation in *Biochemical Individuality*	Page(s) of Citation in *Biochemical Individuality*	Other Publications Containing the Identical Citation	Page(s) of Citation in Other Publications
25	T. S. Danowski, Shirley Hedenburg, and Jean H. Greenman, J. Clin. Endocrinol., 9, 768–773, 1949.	53	Biochemical Genetics and its Human Implications (1956)	166
26	Evelyn B. Man and Edwin F. Gildea, J. Biol. Chem., 119, 769–780, 1937.	54	Biochemical Genetics and its Human Implications (1956)	166
27	Ralph E. Bernstein, J. Lab. Clin. Med., 40, 707–717, 1952.	60	Biochemical Genetics and its Human Implications (1956)	167
28	Arnold E. Osterberg, Frances R. Vanzant, Walter C. Alvarez, and Andrew B. Rivers, Am. J. Digestive Diseases, 3, 35–41, 1936.	60	Biochemical Genetics and its Human Implications (1956)	167
29	Pauline Beery Mack, Personal Communication.	64	Biochemical Genetics and its Human Implications (1956)	167
30	F. R. Steggerda and H. M. Mitchell, J. Nutrition, 31, 407–422, 1946.	137	Biochemical Genetics and its Human Implications (1956)	170
			Human Nutrition and Individual Variability (1956)	12, 13
31	William C. Rose, Federation Proc. 8, 546–552, 1949.	141	Biochemical Genetics and its Human Implications (1956)	170
			Human Nutrition and Individual Variability (1956)	20

(*continued*)

Tab. 3: Continued

No.	Citation in *Biochemical Individuality*	Page(s) of Citation in *Biochemical Individuality*	Other Publications Containing the Identical Citation	Page(s) of Citation in Other Publications
32	William C. Boyd, Genetics and the Races of Man, Little Brown & Co., Boston, Mass., 1950.	49, 127	Biochemical Genetics and its Human Implications (1956)	170
33	Arthur L. Fox, Personal Communication.	128	Biochemical Genetics and its Human Implications (1956)	171
34	R. L. Kirk, and N. S. Stenhouse, Nature, 171, 698–699, 1953.	128	Biochemical Genetics and its Human Implications (1956)	171
35	A. J. Clark, The Mode of Action of Drugs on Cells, Edward Arnold & Co., London, England, 1953.	107, 110	Biochemical Genetics and its Human Implications (1956)	171
36	Jane E. Denton and Henry K. Beecher, J. Am. Med. Assoc., 141, 1051–1057, 1146–1153, 1949.	128	Biochemical Genetics and its Human Implications (1956)	171
37	Recommended Dietary Allowances, National Academy of Sciences, National Research Council, Washington, D.C., Publication 302 (rev.), 1953, p. 10.	137, 143, 148, 151, 153, 155, 156	Human Nutrition and Individual Variability (1956)	12, 16, 19
38	Erich Bloch, Ph.D. Thesis, The University of Texas, Austin, 1953.	138	Human Nutrition and Individual Variability (1956)	13
39	F. E. Lovelace, C. H. Liu, and C. M. McCay, Arch. Biochem., 27, 48–56, 1950.	138	Human Nutrition and Individual Variability (1956)	13
40	Henry E. Paul and Mary F. Paul, J. Nutrition, 31, 67–78 (1946).	144	Human Nutrition and Individual Variability (1956)	13

Biochemical Individuality 201

Tab. 3: Continued

No.	Citation in *Biochemical Individuality*	Page(s) of Citation in *Biochemical Individuality*	Other Publications Containing the Identical Citation	Page(s) of Citation in Other Publications
41	Hans Popper and Frederick Steigmann, J. Am. Med. Assoc., 123, 1108–1114, 1943.	145	Human Nutrition and Individual Variability (1956)	15
42	Tom D. Spies and Hugh R. Butt, "Vitamins and Avitaminoses," in Garfield G. Duncan, ed., Diseases of Metabolism, p. 520.	147, 155	Human Nutrition and Individual Variability (1956)	15, 19
43	Fuller Albright. Allan M. Butler, and Esther Bloomberg, Am. J. Diseases Children, 54, 529–547, 1937.	146	Human Nutrition and Individual Variability (1956)	16
44	C. I. Reed, H. C. Struck, and I. E. Steck, Vitamin D, University of Chicago Press, Chicago, Ill., 1939.	147	Human Nutrition and Individual Variability (1956)	16
45	Oliver H. Lowry, Physiol. Revs., 32, 431–448, 1952.	140, 148, 150	Human Nutrition and Individual Variability (1956)	16
46	Susan B. Merrow, R. F. Krause, J. H. Browe, C. A. Newhall, and H. B. Pierce, J. Nutrition, 46, 445–458, 1952.	148	Human Nutrition and Individual Variability (1956)	16
47	Alice B. Kline and Mary S. Eheart, J. Nutrition, 28, 413–419, 1944.	148	Human Nutrition and Individual Variability (1956)	16
48	Gilbert Dalldorf, "Vitamin C in Health and Disease," in Michael G. Wohl, ed., Dietotherapy, W. B. Saunders Co., Philadelphia, Pa. and London, England, 1945, pp. 293–305.	147, 149	Human Nutrition and Individual Variability (1956)	17
49	W. F. Lamoreux and F. B. Hutt, J. Agr. Research, 58, 307–316, 1939.		Human Nutrition and Individual Variability (1956)	17

(continued)

Tab. 3: Continued

No.	Citation in *Biochemical Individuality*	Page(s) of Citation in *Biochemical Individuality*	Other Publications Containing the Identical Citation	Page(s) of Citation in Other Publications
50	Victor A. Najjar and L. Emmett Holt, Jr., J. Am. Med. Assoc., 123, 683–684, 1943.	150	Human Nutrition and Individual Variability (1956)	17
51	Hellin A. Louhi, Hsi-Hsuan Yu, Betty E. Hawthorne, and Ciara A. Storvick, J. Nutrition, 48, 297–306, 1952.	152	Human Nutrition and Individual Variability (1956)	18
52	Otto Bessey, Personal Communication.	152	Human Nutrition and Individual Variability (1956)	14, 18
53	L. B. Pett, Can. J. Pub. Health, 36, 69–73, 1945.	152	Human Nutrition and Individual Variability (1956)	19
54	D. V. Tappan, U. J. Lewis, U. D. Register, and C. A. Elvehjem, J. Nutrition, 46, 75–85, 1952.	153	Human Nutrition and Individual Variability (1956)	19
55	A. Hansen and O. Bessey, Personal Communication.	155	Human Nutrition and Individual Variability (1956)	19
56	Norman C. Wetzel, Howard H. Hopwood. Manuel E. Kuechle, and Robert M. Grueninger, J. Clin. Nutrition, 1, 17–31, 1952.	156	Human Nutrition and Individual Variability (1956)	19
57	Norman C. Wetzel, Warren C. Fargo, ami Isabel H. Smith, Science, 110, 651–653, 1949.	156	Human Nutrition and Individual Variability (1956)	19
58	Walter G. Unglaub, Harold L. Rosenthal, and Grace A. Goldsmith, J. Lab. Clin. Med., 43, 143–156, 1954.	156	Human Nutrition and Individual Variability (1956)	19
59	R. W. Engel, Proc. Soc. Exptl. Biol. Med., 50, 193–196, 1942.	156	Human Nutrition and Individual Variability (1956)	19

Tab. 3: Continued

No.	Citation in *Biochemical Individuality*	Page(s) of Citation in *Biochemical Individuality*	Other Publications Containing the Identical Citation	Page(s) of Citation in Other Publications
60	D. H. Copeland, Proc. Soc. Exptl. Biol. Med. 57, 33–35, 1944.	156	Human Nutrition and Individual Variability (1956)	19
61	O. M. Hale and A. E. Schaefer, Proc. Soc. Exptl. Biol. Med., 77, 633–636, 1951.	156	Human Nutrition and Individual Variability (1956)	19
62	Helen Kirby Berry, Univ. Texas Publ., 5109, 157–164, 1951.	59	Human Nutrition and Individual Variability (1956)	20
63	Anthony A. Albanese, Protein and Amino Acid Requirements of Mammals, Academy Press, New York, N.Y., 1950.	142	Human Nutrition and Individual Variability (1956)	20
64	F. William Sunderman and F. Boerner, Normal Values in Clinical Medicine, W. B. Saunders Co., Philadelphia, Pa. and London, England, 1949.	52, 53, 135	Human Nutrition and Individual Variability (1956)	20
65	M. P. Hutt, Am. J. Med. Sci., 223, 179, 1952.	135	Human Nutrition and Individual Variability (1956)	20
66	Abraham Cantarow, "Mineral Metabolism," in Garfield G. Duncan, Diseases of Metabolism, W. B. Saunders Co., Philadelphia, Pa., 3rd ed., 1953, pp. 237–213.	136	Human Nutrition and Individual Variability (1956)	20

12.1 Evidence

The writing of this book is based upon the need in human biology and medicine for more attention to variability and individuality at the physiological and biochemical levels. (…) No attempt to bring together the available biochemical material or normal variation has been previously made so far as I know. (Williams, 1956a, p. ix)

The opening words to *Biochemical Individuality: The Basis for the Genetotrophic Concept* set out its raison d'être: collecting data and evidence for the existence of variation amongst human beings. A secondary goal is "get[ing] scientists in general to see enough importance in physiological variability, so that some substantial research will be done on the subject (...)" (Williams, 1955b). Citing the often-anecdotal nature of this evidence, *Biochemical Individuality* sets out to prove that the study of individuality is more than just a "hobby" (Williams, 1956a, p. X). With large portions of the available data described as "far from being satisfactory", it is yet another call to arms for society to appreciate and study the inherent differences between individuals (Williams, 1956a, p. X).

Comprising fourteen chapters and two-hundred and nine pages, *Biochemical Individuality* entails all aspects of BI, providing a plethora of examples to reinforce the validity of the theories described. The importance of providing this evidence is highlighted by the substantial disclosure of references at the end of each chapter (with the exception of Chapter XII), the entire book containing a total of four-hundred and seventy-eight citations (Williams, 1956a, pp. 6–7, 17, 45–46, 66–68, 78–79, 95–96, 104–105, 117–118, 132–134, 162–165, 176, 195–196, 208–209). Ninety of these citations can be found within the bibliographies of Williams' prior publications (excluding those from 1956), indicating that *Biochemical Individuality* is the culmination and further development of the research of preceding decades.[333] Differing from previous books, such as *The Human Frontier*, anecdotal evidence is kept to a minimum, with concrete values, images, and tables at the forefront.[334] The entire book contains 40 tables, figures, and graphs, all of which are based upon evidence collected by other scientists and within DBUT (Williams, 1956a, pp. 19, 21, 22, 23, 24, 25, 26, 27, 29, 31, 33, 34, 36, 37, 38, 41, 42, 50, 51, 52, 55, 56, 59, 60, 61, 73, 76, 87, 92, 98, 100, 102, 113, 120, 146, 152, 170, 188). Having specified and sharpened the concept of BI in the ten years following the publication of *The Human Frontier*, a more scientific and evidence-based approach is chosen in *Biochemical Individuality*. The largest proportion of the cited materials in *Biochemical Individuality* are published after 1946, indicating that the evidence for individuality largely proliferates in this

---

333 See Tab. 2: Articles Cited by *Biochemical Individuality* Appearing in Prior Publications. References 1, 18, 21, 28, 33, 52, 53, 55, 72, and 75 are cited in more than one chapter and therefore constitute multiple citations in *Biochemical Individuality*. These have been counted accordingly.
334 See Chapter 6. *The Human Frontier*.

timeframe (Williams, 1956a, pp. 6–7, 17, 45–46, 66–68, 78–79, 95–96, 104–105, 117–118, 132–134, 162–165, 176, 195–196, 208–209). Twenty-two of Williams' own publications are cited within the book on twenty-nine occasions, indicating the essential nature of his own research to the final product.[335] The earliest of his own publications to share citations with *Biochemical Individuality* is *The Human Frontier*, further highlighting the importance of this publication to the development of BI. In total, just under one-fourth of the references in *Biochemical Individuality* are either written or used by Williams in the decades before its publication, further underlining its status as the crystallisation point and summation of all preceding research on BI.[336]

Tab. 4: Citations of Williams' Own Publications in *Biochemical Individuality*

No.	Citation in *Biochemical Individuality*	Page(s) of Citation in *Biochemical Individuality*
1	Roger J. Williams, William Duane Brown, and Robert W. Shideler, *Proc. Natl. Acad. Sci. U.S.*, **41**, 615–620 (1955)	3, 188
2	R. J. Williams, V. H. Cheldelin, and H. K. Mitchell, *Univ. Texas Publ.*, **4237**, 97–104 (1942).	62
3	Roger J. Williams, *Quart. J. Studies Alc.*, 7, 567–587 (1947).	108, 159, 184
4	R. J. Williams, *Science*, **74**, 597–598 (1931).	126
5	Roger J. Williams, *Univ. Texas Publ.*, **5109**, 10–12 (1951).	128
6	Roger J. Williams, *The Human Frontier*, Harcourt, Brace & Company, New York, N.Y., 1946, pp. 71–72, 73–76.	128, 130, 154
7	Roger J. Williams, "The Chemistry and Biochemistry of Pantothenic Acid," *Advances in Enzymology*, Interscience Publishers, Inc., New York, N.Y., 1943, Vol. III, pp. 253–287.	154
8	Esmond E. Snell, Derrol Pennington, and Roger J. Williams, *J. Biol. Chem.*, **133**, 559–565 (1940).	154
9	Roger J. Williams, "The Clinical Possibilities of Pantothenic Acid," in Michael G. Wohl, ed., *Dietotherapy*, pp. 263–267.	154

(*continued*)

---

335 See Tab. 4: Citations of Williams' Own Publications in *Biochemical Individuality*. References 1, 3, 6, 10, 11, and 15 are cited in more than one chapter and therefore constitute multiple citations in *Biochemical Individuality*. These have been counted accordingly.

336 $(90+29)/478 = 0{,}249$.

Tab. 4: Continued

No.	Citation in *Biochemical Individuality*	Page(s) of Citation in *Biochemical Individuality*
10	Roger J. Williams, Richard B. Pelton, and Lorene L. Rogers, *Quart. J. Studies Alc.*, **16**, 234–244 (1955)	161, 185
11	Roger J. Williams, L. Joe Berry, and Ernest Beerstecher, Jr., *Arch. Biochem.*, **23**, 275–290 (1949).	173, 185
12	Roger J. Williams, Ernest Beerstecher, Jr., and L. Joe Berry, *The Lancet*, February 18, 1950, pp. 287–294.	174
13	R. J. Williams, *Nutrition Revs.*, **8**, 257–260 (1950).	174
14	Roger J. Williams and Lorene L. Rogers, *Texas Repts. Biol. Med.*, **11**, 576 (1953).	175
15	Roger J. Williams, *Free and Unequal*, University of Texas Press, Austin, Tex., 1953.	176, 198
16	Roger J. Williams, *J. Public Law*, **3**, 328–344 (1955).	176
17	Ernest Beersterher, Jr., H. Eldon Sutton, Helen Kirby Berry, William Duane Brown, Janet Reed, Gene B. Rich, L. Joe Berry, and Roger J. Williams, *Arch. Biochem.*, **29**, 27–40 (1950)	184
18	Unpublished findings, Biochemical Institute, The University of Texas.	184
19	Roger J. Williams, L. Joe Berry, and Ernest Beerstecher, Jr., *Proc. Natl. Acad. Sci. U.S.*, **35**, 265–271 (1949).	185
20	Roger J. Williams, *J. Clin. Nutrition*, **1**, 32–36 (1952).	185
21	Roger J. Williams, *Nutrition and Alcoholism*, University of Oklahoma Press, Norman, Okla., 1951.	185
22	Lorene L. Rogers, Richard B. Pelton, and Roger J. Williams, *J. Biol. Chem.*, **214**, 503–506 (1955).	185

12.1.1 From Basic Genetics to All-Encompassing Variation

Human genetics and the principles derived therefrom form a base of understanding on which the comprehension of Biochemical Individuality stands. With genetic differences creating ubiquitous variation, Williams explains the genetic principles relevant to BI in Chapter II. Partial genetic blocks and their relevance are highlighted once more, while the lack of appreciation for the importance of genetics regarding disease is additionally discussed (Williams, 1956a, p. 10).[337]

---

337 See Section 8.1. The Metabolic Individualities of Rats.

The interplay of heredity and environment is clarified in "a plea for an unprejudiced facing of the facts of heredity" (Williams, 1956a, p. 16). Critiquing the often-fatalistic view of heredity, Williams argues that BI and GC offer an answer to many unsolved questions.

Laypeople its primary target, Williams begins his elucidations with the most obvious examples of human differences in *Biochemical Individuality*. Using Anson's *Atlas of Human Anatomy* as a primary source (Anson, 1950), the manifold variations of anatomical minutia are discussed in greater detail than in all prior publications.[338] Variations of a wide range of organ systems and their functions/activity are depicted, encompassing "brain, nerves, muscles, tendons, bones, blood, organ weights, [and] endocrine (…)" (Williams, 1956a, p. 45, 1956a, chap. III, VI, VII). This provides an illustrative overview of how diverse human beings can be beyond their superficial visible variations. Not limited to macroscopic differences, the microscopic disparities in chemical make-up of blood, bodily fluids, and structures are also demonstrated (Williams, 1956a, chap. IV). Similarly, the assorted enzymic variations shown by individuals is described in great detail, with singular enzymes and their exact values of variation portrayed (Williams, 1956a, chap. V). Variations regarding pharmacological reactions, one of the examples essential to Williams' first notions of individuality,[339] and miscellaneous examples of individuality, including the first example publicised by Williams in 1931 (Williams, 1956a, pp. 127–130),[340] are also given due consideration (Williams, 1956a, chap. VIII, IX). Nutritional individuality is the final subject for which Williams provides a large number of examples (Williams, 1956a, chap. X). All of these subjects are reflected in some way by Williams' prior publications, yet *Biochemical Individuality* goes into much greater detail and furnishes an enormous amount of statistical data to impress the scope and importance of these variations. In this sense, a considerable amount of new and previously unseen data is presented. This further indicates how *Biochemical Individuality* largely widens the scope of BI as a continuation of decades of research.

12.2 Variation and Its Significance

The names Darwin, Hippocrates, and Galen are widely known throughout society, regardless of vocational training or scientific expertise. It is with references to these familiar individuals that Williams opens *Biochemical Individuality*, drawing

---

338   See Section 11.3. Individual Anatomies.
339   See Section 4.3. Adverse Drug Reaction.
340   See Section 4.5. "'Taste Deficiency' for Creatine".

interest from the reader and legitimising his following claims. Quotations depicting the individuality of human beings dating back thousands of years indicate the longevity, and therefore relevance, of the study of individuals, while simultaneously providing a backdrop for further thought. Paired with details of his own research, and the evidence for individuality it provides, a strong case is made for the importance and potential implications of Williams' claims. While expanding the collection of data indicating the existence of Biochemical Individuality may further impress its relevance on the reader, this does not carry much weight regarding the further development of the theory. Similarly, the chapter expounding on the Genetotrophic Principle heavily relies on Williams' previous publications (Williams, 1956a, chap. XI). *Biochemical Individuality* signifies such development in its final three chapters, in which Williams discusses the possible implications of his theory.

Widely critiquing scientists' tendency to ignore variation and individuality in most of his previous research with only modest success, Williams' approach is altered in *Biochemical Individuality*.[341] While repeating his assertion that disregarding variation can produce only inaccurate results, this statement is restricted to the application of knowledge from biological research (Williams, 1956a, p. 177). In basic biological science, variation is described as an enemy (Williams, 1956a, pp. 178–179). Studies must be designed to produce as little variation as possible, as their objective is a fundamental understanding. When such principles are, however, duly understood, their application cannot be implemented without an appreciation for the individuality of the subjects they are being applied to (Williams, 1956a, pp. 178–179). Vision is presented as an ideal example for this line of thinking: the principles of sight can be understood without correcting for wide variation within the human population. When applying this knowledge to improving human vision, however, ignoring an individual's particular needs would lead to the furnishing of worthless visual aids (Williams, 1956a, p. 178), in a sense forcing humanity to accept and correct for variation (Williams, 1956a, p. 178).

It is this "differentiation between the 'pure science' of biology and the 'applied' science of biology" in which *Biochemical Individuality* represents progress within the development of BI (Williams, 1956a, p. 181). On previous occasions, Williams demands that all aspects of the scientific process must incorporate individuality, therefore adding great complexity to such research. Here, he accedes

---

341   See Section 9.3. *Nutrition and Alcoholism* and Chapters 6. *The Human Frontier*, 10. *Free and Unequal*, and 11. Practical Genetotrophism.

that individuality may be relevant to certain aspects of biological research, yet entirely lacking in implications for others (Williams, 1956a, p. 177). Such honing and specifying is due to Williams' realisation that wide acceptance of his theories is only possible if the scientific community does not see BI as a complicating factor (Hodge and Williams, 1980a). While so implied by his prior statements, *Biochemical Individuality* attempts to rectify this issue and make real-world application possible.

The term "humanics" and Williams' proposed science of mankind are an essential steppingstone within the context of BI's development. A central theme in a large number of publications, the study of humankind as part of a new scientific approach is previously depicted as vital to its appreciation. Though the term ceases to be used regularly many years before *Biochemical Individuality* is published, no preceding publication offers an explanation therefore. Due to the "tremendous array of items in which human individuality is exhibited", such an approach is deemed impractical by 1956 (Williams, 1956a, p. 183).

A new approach is suggested, "select[ing] specific problems for study and then *investigat*[ing] *how individual human differences enter into these specific problems*" (Williams, 1956a, p. 183). This method, considered more feasible, is to be based on a three-step formula regarding medical and dental research. Following the selection of a disease with an unknown aetiology, possibly related established and unidentified metabolic individualities are to be sought out. Finally, the Genetotrophic Principle may be applied in order to treat or prevent said disease (Williams, 1956a, p. 185). Discussions on these grounds are lead for gout, arthritis, alcoholism, dental caries, and the common cold (Williams, 1956a, chap. XIII). Finally, the case for increased individualisation in medicine is made, criticising the "assembly line" approach of medicine at the time (Williams, 1956a, p. 195).[342]

Williams pleads a similar case for the study and potential treatment of psychiatric disorders, for which he wishes to furnish an additional approach (Williams, 1956a, pp. 198, 207). Afflictions of the psyche, as well as a tendency toward such, are suggested to also be affected by metabolic individuality and morphological differences in brain structure (Williams, 1956a, p. 198). As with any other organ, the metabolism of the brain is described as entirely individual. Utilising the same approach as discussed above, Williams hypothesises the significance of these deviations may be revealed (Williams, 1956a, pp. 199–200).

---

342   See Section 12.4. Reviews of *Biochemical Individuality*.

This rethinking of the necessary approach to study individuality is a significant deviation from previous publications and denotes further development of the concept of BI. Earlier theories meet the limitations and difficulties involved with real-world study, necessitating a correction in course. This is true for all studies pertaining to BI, not merely medical, psychiatric, and dental research. Abandoning an unrealistically complex and expensive approach, Williams' new advances attempt to utilise existing structures and integrate BI and GC into these.

12.3 Resistance and Publishing Difficulties

Though Williams is elected to become the 1957 president of The American Chemical Society in 1955, indicating his considerable renown within the scientific community, the publication process of *Biochemical Individuality* presents an unexpected challenge (Dr. Williams, '57, President of ACS, 1955; Texas Professor Heads American Chemical Group, 1955; Texas Professor Honored as Head Of Chemist Group, 1955; University of Texas Prof is Named New President of ACS, 1955; US Chemists Name UT Prof President, 1955; UT Biochemist To Head Society, 1955; UT Professor Elected Head American Chemical Society, 1955). Though he is described most favourably as having an "impressive background in biochemistry" (Bennett, 1958), as "one of the most distinguished biochemists of our time" (Greenstein, 1957), and "productive, experienced and widely known" in reviews of *Biochemical Individuality* (Grenell, 1957), the process of publishing the book is an arduous one. In letters to Warren Weaver, a regular correspondent and confidant, Williams indicates a remarkable level of hesitance to publish *Biochemical Individuality* by various publishing houses (Williams, 1955b).[343]

This hesitance is, so Williams, not based upon a lack of agreement with or quality of his material, but "medical and scientific bigotry" (Williams, 1955b). With far-reaching medical implications contained in the book, Williams' theories do not necessarily follow the medical trends of the time. The leading contemporary medical vision is described to be diametrically opposed to Williams' views, as "with the advent of improved diagnostic tools for detecting and characterizing disease and new drug interventions that were working apparent clinical miracles uniformly across population groups, conventional medicine was rapidly moving to standardization of diagnosis, of therapy, and, well, of people" (Pizzorno, 2019, p. 1). Indicating variation at a cellular level, BI goes stringently against this stream of though. Citing such "real resistance" to his ideas, Williams shows the importance he attributes to BI by re-submitting his work until it is

---

343   See Section 3.2. Academic Networks.

finally published (Williams, 1955c). Seriously considering the initiation of his own journal to publish work on individuality, a pronounced frustration with the process of publication becomes apparent (Williams, 1955b). With a first manuscript of *Biochemical Individuality* finished in late 1954 (Williams, 1954c), the book is ultimately published by John Wiley & Sons, Inc. in 1956, following a two-year process (Williams, 1956a).[344]

12.4 Reviews of *Biochemical Individuality*

Following the difficulties and described resistance involved with publishing *Biochemical Individuality*, the reviews of said book paint a much more positive picture than may be expected.[345] The general tenor of all reviews found at the time of writing is positive, with the facts as laid out by Roger Williams commended as cogent and well-presented by doctors, psychiatrists, biologists, geneticists, and (bio)chemists alike (Bean, 1957; Beloff, 1957; Bennett, 1958; Biochemical Individuality. The Basis for the Genetotrophic Concept, 1957; Boyd, 1957; Candlish, 1989; Greenstein, 1957; Grenell, 1957; Harris, 1958; Huxley, 1957; Kandutsch, 1957; Kety, 1957; Keys, 1957; King, 1957; Livermore, 1958; McCay, 1957; Paterson, 1957; Pauling, 1957; Pope, 1958; Williams, 1957d). The general thesis and importance attributed to Biochemical Individuality is also accepted by reviewers. It is highly commended as the first book of its kind, attempting to collect as much information as possible on the individualities shown by human beings (Beloff, 1957; Boyd, 1957; Greenstein, 1957; Kandutsch, 1957; Keys, 1957). A key difference to Williams' two other more extensive books on BI is that *Biochemical Individuality* deals with purely scientific matters.[346] Discussing issues largely outside of Williams' field of expertise, his other two books were widely criticised.[347] His high regard in all matters concerning biochemistry and science, therefore, provides a staple of quality and validity which the other books lacked. The book is also less dogmatic in its essence. While calling for complete societal overhaul in previous publications, *Biochemical Individuality* concentrates only on the scientific implications of BI.[348]

---

344 John Wiley & Sons, Inc. is one the leading publishers of scientific and technical material. It was founded in 1807 and is more colloquially known as Wiley. For further information, see Wiley (2023).
345 See Section 12.3. Resistance and Publishing Difficulties.
346 See Chapters 6. *The Human Frontier* and 10. *Free and Unequal*.
347 See Sections 6.6. Reviews of *The Human Frontier* and 10.4. Reviews of *Free and Unequal*.
348 See Chapters 6. *The Human Frontier* and 10. *Free and Unequal*.

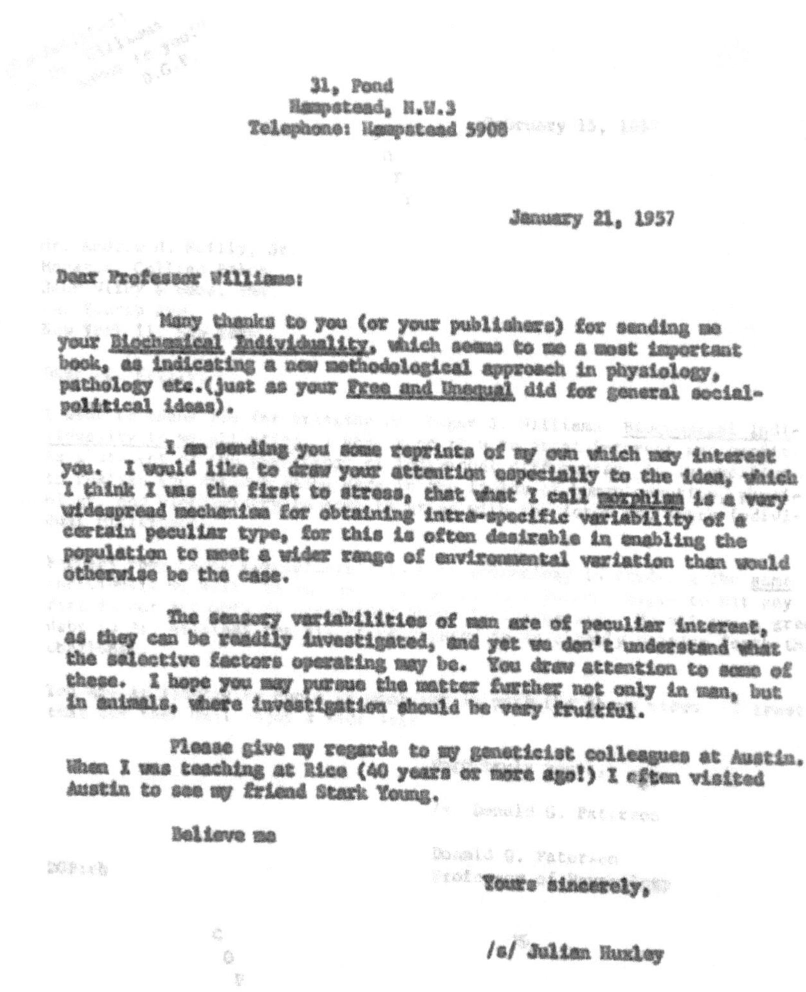

> 31, Pond
> Hampstead, N.W.3
> Telephone: Hampstead 5908
>
> January 21, 1957
>
> Dear Professor Williams:
>
> Many thanks to you (or your publishers) for sending me your Biochemical Individuality, which seems to me a most important book, as indicating a new methodological approach in physiology, pathology etc.(just as your Free and Unequal did for general social-political ideas).
>
> I am sending you some reprints of my own which may interest you. I would like to draw your attention especially to the idea, which I think I was the first to stress, that what I call morphism is a very widespread mechanism for obtaining intra-specific variability of a certain peculiar type, for this is often desirable in enabling the population to meet a wider range of environmental variation than would otherwise be the case.
>
> The sensory variabilities of man are of peculiar interest, as they can be readily investigated, and yet we don't understand what the selective factors operating may be. You draw attention to some of these. I hope you may pursue the matter further not only in man, but in animals, where investigation should be very fruitful.
>
> Please give my regards to my geneticist colleagues at Austin. When I was teaching at Rice (40 years or more ago!) I often visited Austin to see my friend Stark Young.
>
> Believe me
>
> Yours sincerely,
>
> /s/ Julian Huxley

Fig. 19: Carbon Copy of a Letter by Julian Huxley to Williams on 21.01.1957 indicating his interest and appreciation for the contents of *Biochemical Individuality*; Dolph Briscoe Center for American History, The University of Texas at Austin, Roger Williams Papers, Box 88-087/26a, Folder: Famous Names (Humanics).

Though their general tenor is ubiquitously positive, the reviews are not without any criticism. Two reviewers criticise the lack of completeness regarding evidence for individuality (Grenell, 1957; McCay, 1957), while one reviewer objects to two specific examples of individuality presented (Boyd, 1957).[349] Others challenge whether the book really presents any truly revolutionary thought to their respective fields (Kandutsch, 1957; McCay, 1957). All in all, however, the reception of *Biochemical Individuality* is remarkably positive, especially when compared to Williams' previous publications.[350] Described as "clear and lucid" and "thought-provoking", a review in *The Yale Journal of Biology and Medicine* recommends his work to all "in the field of medicine and psychiatry, but also to students in all phases of biological science, particularly to those in applied biology" (Beloff, 1957). This opinion is also shared by personal correspondence following its publication (Huxley, 1957; Paterson, 1957; Williams, 1957d).

The reviews of *Biochemical Individuality* additionally provide information on the scientific and medical developments during the years in which Williams' theories mature. Highlighting the advances in (bio)chemical methods for analysing metabolism, some of which Williams himself played an important role in developing (Kirby Berry et al., 1951), one review discusses how much of the information cited would be unobtainable without the technical developments within the last generation (Kety, 1957).[351] A medical review highlights the change in focus of medical doctors when treating and describing disease within the 20th century. With the "vast proliferation of information", and the discovery of more and more diseases, the development of nosology as a medical science becomes necessary (Bean, 1957).[352] Rather than focusing attention on the specific patient afflicted with an illness, with much "thought and action based on ideas of diathesis, disposition and temperament", the increased stock of knowledge makes such focused efforts impractical (Bean, 1957). Pragmatism diverting the clinician's eye from the individual before them, the importance of individual variation, in some aspects central to medicine historically, is diminished (Bean, 1957).

---

349 Boyd (1957, pp. 141–142) criticises Williams' loose use of the term "allergy" and the fact that creatine does not taste bitter to any individual known to him.
350 See Sections 6.6. Reviews of The Human Frontier, 9.3.4. Promotion and Reviews of *Nutrition and Alcoholism*, and 10.4. Reviews of *Free and Unequal*.
351 See Sections 5.2. The Vitamin Content of Tissues and 9.1. Biochemical Individuality V.
352 Nosology is defined as "a branch of medical science that deals with classification of diseases" by Merriam-Webster (2023).

12.5 Appreciation Through the Years

The strikingly positive reception of *Biochemical Individuality* in the years following its publication indicates its importance and relevance. As an overview of the scope of interindividual variations, it is still appreciated more than 65 years later. It was reissued in 1998 and is still available for purchase (Williams, 1998). Similarly, further books written by Williams have been and reissued following their original publication (Williams, 1979, 2018). Quoted in works even decades after its publications, *Biochemical Individuality* is still seen by many as a standard work within the context of human individuality (Candlish, 1989; Gonzalez and Massari, 2012; Motulsky, 2002; Neustadt and Pieczenik, 2007; Pizzorno, 2019).[353]

12.6 Biochemical Individuality in Its Final Form

Every human being is an entirely unique individual. Be it macroscopic differences in anatomy or microscopic and biochemical variances, every single aspect of an individual differs from his or her peers. These differences have been proven to exist in studies carried out by scientists of multiple fields and may be the basis of many diseases of obscure origin. Many of these diseases may be genetotrophic in nature and could therefore be successfully treated by nutritional supplementation. Though fundamental scientific research may disregard variation in order to foster understanding and progress, the appreciation and study of these individualities is necessary for the real-life application of this knowledge. Applying the knowledge gained through the study of individuals can be the basis of solving countless issues facing society.

12.7 Conclusion

Biochemical Individuality is Roger Williams' most important publication on the topic of individuality. A crystallising point for multiple decades of research, this book summarises all aspects of BI explored within the previous years and develops these further. Expanding the volume of evidence for a wide variety of examples indicating interindividual human individuality, it builds upon the citations and examples presented in the previous publications discussed above. Discussing possible implications for biological, medical, dental, and psychiatric research, it cements the relevance of Biochemical Individuality and the Genetotrophic Principle with irrevocable data. Received well by critics and colleagues alike, it is still celebrated today as the first large collection of data on individuality.

---

353   See also Section 1.2. Literature Review.

13 Results

Following the above accumulation of evidence toward the three guiding questions posed at the outset of this analysis, this chapter seeks to condense the central implications of the material presented above. Illustrating the results of the aforementioned studies, it will discuss the facts and extrapolations therefrom as pertinent to the development of BI as a concept, the relevance of external individuals, scientific material, and the historical context, as well as the role of *Biochemical Individuality* as a crystallisation point of Williams' research. To this end, said evidence will be presented in chronological order, beginning with childhood musings, and ending with the publication of *Biochemical Individuality* in 1956. The underlying significance of these facts will then be discussed in the following chapter.

The cumulative presented evidence suggests that Williams' concept of Biochemical Individuality first arose haphazardly and coincidentally alongside other research through personal encounters. Originating in its most basic form from childhood-experiences, Williams' first ideas of BI have been shown to be obscure. They reflect every human's inherent understanding of their own individuality from a young age, as well as the predominant understanding of heredity in the late 19th century. The available literature has indicated how thoughts of studying such individuality are first overshadowed by a conquest for the discovery of new vitamins at the beginning of Williams' career. Such concrete anecdotal evidence is provided by Williams' personal experience with an idiosyncratic reaction to morphine, sparking interest in human pharmacological variation. With a first notion of individuality in a medical and biochemical sense documented through Williams' discussion of "individual metabolic idiosyncrasies" in an article entitled " 'Taste Deficiency' for Creatine" in 1931, what may seem like unrelated studies using yeast and fungi offer evidence and understanding for BI in the following years, though they are not immediately attributed this relevance. In hindsight, Williams additionally attributes importance to his dissertational research, which provides data on variation in yeast. These aforementioned examples constitute first steps toward the development of the concept of BI, though they are more speculative and less discernible than later progress.

Alongside a growing stockpile of biochemical knowledge and ongoing large-scale investigations, further evidence of individuality in yeast strains becomes apparent through the analysis of publications related to pantothenic acid. With the existence of variation in single-celled organisms established there through, the distinctiveness of more complex species has been shown to be the focus of Williams' following publications. Articles connected to the study of the vitamin content of chick tissues additionally provide first tentative hints at an appreciation for individuality. These include first allusions to individuality and its potential importance, which become more pronounced in discussions regarding human nutrition in the following years. This underlines the slow and steady process of understanding and appreciation for the extent of individuality in all living organisms. Following the publication of a first monograph considered "public education", *What To Do About Vitamins*, in 1945, Roger Williams' work is increasingly involved in the study of what later becomes the concept of Biochemical Individuality. Subsequent research is characterised by an increasing attention to popularising his theories, as evidenced by a number of publications aimed at the general public. The discovery of individuality at a cellular level and its tracing through increasing tiers of complexity creates the impression of a journey of understanding.

In a further stage of the development of BI, a first publication identified to be entirely dedicated to this concept is released with *The Human Frontier* in 1946 (Williams, 1946a). In this sociological monograph with mixed reviews, Williams collects a variety of examples for the uniqueness of individuals. These range from the anecdotal evidence characterising previous publications to concrete scientific data indicating interindividual variation. An appreciation for the importance of BI becomes evident by this work appealing for the creation of a new branch of science named "humanics". The publication of *The Human Frontier* is a true turning point in both Williams' career and the development of BI, with the subject becoming a concrete and defined matter of interest. In its wake, the conjugate term Biochemical Individuality appears for the first time, and first own scientific research dedicated to individuality becomes evident. Basing much of this publication's understanding on Beadle and Tatum's "one gene – one enzyme" hypothesis, Loeb's research of individuals, Draper's susceptibility studies, and Garrod's deliberations on "inborn errors of metabolism", the importance of external research additionally becomes clear and provides an example for the interconnectedness of Williams' work with preceding scientific research. The need for popular support manifests following largely negative reviews from anthropological and sociological colleagues, leading to the promotion of his work within the public eye. Through the diversification of Williams' theories within the historical context of post-World War Two America, political potentialities become a further aspect in the development of BI. Published efforts to improve the social incorporation of individuals according to their individual talents and capabilities are exemplary therefore. The above, alongside the considerable correspondence found through analysis and comparison of the citations of *The Human Frontier* and *Biochemical Individuality*, allow for the conclusion that *The Human Frontier* can be considered a first precursor publication to *Biochemical Individuality*.

Following first publications indicating ongoing studies of individuality, the term Biochemical Individuality appears in Williams' work as a conjugate term for the first time in 1947, which marks another significant step in the development of the concept. Alongside the study of the origins of alcoholism and its metabolic correlates, an increased attention to the promotion of research efforts leads to a marked diversification in Williams' preferred modes of publication. Strongly influenced by the anti-communist sentiment of the early Cold War and the increased notoriety of scientists following the Manhattan Project, his work additionally presents itself within the contemporary political landscape. With BI becoming an increasingly important factor in Williams' research, considerable numbers of overlapping citations between the publications following *The Human Frontier* and *Biochemical Individuality* further indicate the relevance of his early alcoholism research to the latter publication.

The therapeutic potential of earlier vitamin and nutrition findings have been identified as a primary research focus following first discussions of the Genetotrophic Concept in 1949, certain aspects of which are also reflected in prior publications. Subsequent publications discussing the study of rats through the augmentation of alcohol consumption have indicated an increased conviction for the importance of Williams' discoveries, with multiple joint studies additionally highlighting the collaborative aspects of this research subject. An increase in the relative number of articles discussing individuality additionally emphasises the intensifying importance of BI within the overall context of Williams' research. Greatly diversifying the scope of research, a next tier of complexity in the development of BI is reached by the discussion of GC within the context of "diseases of obscure origin" and further increased by the scientific study of individual human beings. Concrete solutions to the problem of alcoholism are developed and published in another book of public education, *Nutrition and Alcoholism*, which additionally increases Williams' societal notoriety, as evidenced by the large number of letters he receives following this publication.

Publishing large quantities of individuality-related data in *Biochemical Institute Studies IV*, Williams' institute has been shown to contribute to the study of individuals through the development of new techniques for biochemical analysis. Previous published works having discussed the sociological implications of human variation, the political consequences which BI entails, and the structural reform necessary to counteract these, are the focus of a following significant publication. Widely criticising society's tendency toward uniformity in the early stages of the Cold War, BI becomes more dogmatic in its nature, calling for a complete overhaul of many fundamental aspects of society in a monograph entitled *Free and Unequal*. Matters such as sexism, racism, the communist agenda, and biological determinism are subject to in-depth discussion, with the voiced opinions deviating markedly from the scientific and societal trends of the time. With such subjects at the centre of societal discussions in the 21st century, the progressive nature of Williams' works becomes clear. Laying out the facts of individuality as established within the context of BI, the purely data-driven scientific research of individuality previously presented in Williams' publications on the subject is not reflected by this monograph. Similar to previous attempts at discussing topics outside of his vocational training in *The Human Frontier*, the all-encompassing importance attributed by Williams to his concept of BI in the early 1950s becomes evident.

The increasing importance and ongoing development of BI within the overall context of Williams' work is indicated by its central role in his research and publications following the publication of *Free and Unequal*, with an intensification of the collection of concrete and statistically significant data indicated through bibliographical analysis. Once more demonstrating the relevance of external research for Williams' work, Anson's anatomical investigations become a central and evocative source of evidence and are later extensively used in *Biochemical Individuality*. Voted to be the president of the American Chemical Society for 1957, Williams further presents his concept of "chemical anthropology" as part of a speaking tour. Though it largely shares similarities with the previously conceived study of "humanics", and does not reappear in *Biochemical Individuality*, it additionally underlines the universality attributed to BI in Williams' research of the early 1950s. The juxtaposition of Williams' public appearances and his academic publications has here revealed an intention to disseminate the collected knowledge of BI, as is later also reflected in *Biochemical Individuality*.

Biochemical Individuality, published in 1956, encompasses all of Williams' prior knowledge and understanding of BI. Fundamentally scientific in its approach, yet written for laypeople, it is not only a collection of data on individuality but includes a further matured view of BI. Highlighting the importance of BI for a wide variety of scientific fields, the considerable correspondence of citations in *Biochemical Individuality* and almost all publications from 1946 onwards, indicates the cumulative nature of BI. Less dogmatic in its approach, *Biochemical Individuality* produces an irrevocable substantiation of the theories, thoughts, and concepts of the preceding decades as a true crystallisation point of Williams' research. Having previously championed the creation of an entirely new branch of science, *Biochemical Individuality* places the concept's importance at the end of the scientific process: the real-world application of knowledge gained through experimentation and investigation. Appreciated and relevant till this day, it is Williams' most successful and the best-known publication of his career.

14 Discussion

Three guiding questions were posed at the beginning of this analysis, which form the basis for the deliberation and research of Roger Williams' concept of Biochemical Individuality summarised in the previous chapter. Firstly, the above has sought to illustrate the evolution and progression of Williams' concepts concerning Biochemical Individuality between 1919 and 1956 within his publications and personal documents. Furthermore, it has attempted to relate Williams' work to his contemporaries, as well as elucidate the influence of contextual factors on his research. Finally, the relative importance of *Biochemical Individuality: The Basis for the Genetotrophic Concept* as a crystallisation point of Williams' work on individuality was explored. The above results were collected according to these queries, with this discussion now seeking to provide answers through further analysis and reflection. First, the cumulative evidence presented above is analysed in chronological order according to the described queries, followed by a discussion of these results embedded within the previously existing literature. Subsequently, the methodological and circumstantial limitations are discussed followed by the portrayal of an outlook for potential further academic research.

While prior research considering Williams' work on individuality discusses his theories in their final and comprehensive forms, text analysis of Williams' published books and documents between 1919 and 1956 has revealed how his earliest such ideas are unformed and instinctive in nature. These descriptions have provided an indication for how impalpable ideas form a starting point from which his later interest in Biochemical Individuality develops. Further study of contextual scientific knowledge has indicated how the first identified thoughts regarding individuality are shaped by the ideals of Williams' upbringing and the knowledge available at the time. With little to no scientific awareness of genetics and modes of inheritance having been publicly available in the late 19th century, the earliest ideas pertaining to individualism do not go beyond a generic appreciation for macroscopic and concrete palpable differences (e.g., taste, smell, touch, and diverging reactions to the application of medications). Analysis of the text and figures of *Biochemical Individuality: The Basis for the Genetotrophic Concept* have attempted to illustrate how this publication draws on such inherent and instinctive perceptions of individuality. Drawing from these results, it has been suggested that these are used to convey the universality thereof and provide the reader with illustrative examples for individuality, though this cannot be determined with absolute certainty.

Furthermore, the above has indicated how such early vague thoughts become more substantial through the analysis of Williams' dissertational research. This development indicates the retrospective nature of Williams' first inklings regarding individuality. Hindsight reflection on anecdotal stories and early research by Williams himself – and retrograde extrapolation therefrom – further have been suggested to result in the first tangible scientific perceptions regarding biological variation. The close reading of "'Taste Deficiency' for Creatine" has implied that conceptualisations surrounding variation first become concrete through academic publication in connection with a scientifically published anecdotal observation. This includes the labelling of phenomenona surrounding human deviations in taste as "individual metabolic idiosyncrasies" therein, signifying the first term used in Williams' work describing human deviation. Further analysis, however, shows that this phrase does not remain regularly used throughout the aforementioned time period, it has therefore been concluded to be less significant than the later "Biochemical Individuality" and to merely denote a precursor term. Pertaining to the relevance of this work to *Biochemical Individuality: The Basis for the Genetotrophic Concept*, the examples discussed therein reappear when their importance is delineated within its preface.

The study of Williams' research publications pertaining to the structure and function of B-vitamins has shown these to provide understanding only in hindsight, with a diverging research focus, as evidenced by archive material and later reflections, hindering immediate further appreciation. Though these studies and the previously discussed inklings of individuality happen as a by-product of Williams' daily professional life, the study of personal documents and otherwise not commonly available interviews has indicated how they planted the seed of thought. Furthermore, works on the historical development of biochemistry have shown how this early research occurs as part of an overall "race" within the budding field of biochemistry to discover the structure of vitamins and their functions. These studies can, therefore, by nature be described as pioneering and delve into previously uncharted territory. Such first scientific strides are characterised by the aforementioned unknown, this phenomenon has been revealed to continue in the research methodologies used for the study of BI. The in-depth text and figure analysis of *Biochemical Individuality* has furthermore shown this publication to provide countless amounts of data and hard proofs for human variation. The discussed works on vitamins are principally focused on the production of such data and its analysis. Heavily basing later arguments for the general existence of BI on such "basic" scientific research, the analysis of these publications and their research methodologies have demonstrated how Williams' early works display his academic roots.

Through the direct comparison of bibliographical data and the thematic structure of Williams' publications, it has been found that *The Human Frontier: A New Pathway for Science Toward a Better Understanding of Ourselves* signifies a marked break within the overall context of Williams' publications. Through close reading and the analysis of publisher data and archive material, this monography has been shown to not only address a different target group, namely the general public, but also to delve into subjects that Williams had previously neither gathered any professional expertise, nor shown any interest in. The presented evidence indicates that this work is the first to publicise a collection of examples for human variation, many of which have also been shown to be reflected within *Biochemical Individuality: The Basis for the Genetotrophic Concept*. Close reading has provided proof for the fact that this publication additionally contains first attempts to formulate an evidence-based concept of BI into a definition, describing the way in which the relevance of these notions necessitates the creation of an own branch of science. This is named "humanics" – a term which further text analysis has shown to regularly appear in publications until 1954, when it was abandoned. The significance of this term has been revealed through the analysis of personal correspondence, with discussions pertaining to its ultimate rejection additionally illustrated. While the concept of BI has been identified to be more of a notion and by-product of other work prior to this publication, the analysis of Williams' publications within the selected timeframe suggests how it becomes a primary focus from this point onwards.

Additionally, *The Human Frontier* has been identified as the first point in which the research of contemporary and preceding scientists becomes relevant, providing evidence for the second research question of this thesis. The explicit and extensive discussion of the work of five scientists indicates the importance of external theories for the first more concrete concept of BI. As indicated by the analysis of Williams' personal correspondence, manuscripts sent to a number of scientists and prominent people in society additionally denote a first attempt to spread his ideas. The reception and effect of external comments are indicative of the influence that Williams' academic network as well as the external remarks and critiques have on the development of BI. *The Human Frontier* and *Biochemical Individuality* additionally have been shown to overlap considerably regarding the examples of individuality they present. Bibliographical data analysis shows that about 1/5th of the sources discussed in *The Human Frontier* reappear in *Biochemical Individuality*, exhibiting how the former is used as a basis for the production of the latter. *The Human Frontier* has therefore been identified as the first attempt at a collection of examples of individuality, which is later eclipsed by the "ultimate" and extended collection presented in *Biochemical Individuality*, partially answering the third research question.

The overall analysis of Williams' publications has been indicative of the trailblazing nature of the research of alcoholism following the publication of *The Human Frontier*. Close reading has revealed that the conjugate term under which his ideas are later subsumed primarily appears in 1947: Biochemical Individuality. This term has been shown to not only provide a pretext under which scientific research and discussion can be lead, but also aid the promotion of Williams' ideas outside of the scientific community. Though multiple other, less specific and all-encompassing, terminologies are previously discussed, and text analysis has shown these to be largely synonymous with BI, this unification under a fixed expression provides a further element of structure to the previous relative disarray of unattached ideas. Within this study of alcoholism is also contained a transformation in research methods. While studying the research methodologies of Williams' early scientific publications has shown these to create a maximum of homogeneity, the comparison of these to the study of alcoholism has indicated how this later work attempts to achieve the opposite.

The formulation of the Genetotrophic Concept can be described as a natural scientific next step within the overall context of BI. Focusing on individuality within vitamins/nutrition a previously identified field of expertise, the textanalysis of consecutive publications has established that these works entirely focus on the nutritional therapy of diseases arising from individual nutritional needs. GC has been proven to later be a central aspect of *Biochemical Individuality* through the text analysis of this publication. As the full title of this book suggests, the evidence presented therein aims to prove the validity of this concept. The changes to this concept have been determined to be minimal in the following years through the comparative study of manuscripts discussing this topic, while additionally juxtaposing the contents of these works to *Biochemical Individuality*. A methodological shift can here be illustrated from a purely observing role to one involving intervention and active experimentation.

The analysis of personal correspondence has shown how the promotion of his concepts in the public eye increases within the overall context of Williams' alcoholism research, thereby furthering his notoriety and status as an expert on individuality. Through the discussion of the history of alcoholism in the United States, Williams' alcoholism research has been shown to appeal to a wide array of audiences and addresses an issue highly relevant to the American society of the 1940s and 1950s. The large quantity of letters from an extensive variety of societal classes and sectors, which are filed under this subject matter in the

Briscoe Center Archive material, are additionally exemplary of this. Williams' research on alcoholism has, furthermore, been shown to be heavily reflected in *Biochemical Individuality* through bibliographical analysis. This research furthermore provides an aspect of seniority and expertise as a practicing scientist within the field of individuality, as revealed by the comparison of reviews of his publications at various times throughout Williams' career. The fact that BI in turn becomes increasingly important within the overall context of his research following the postulation of the Genetotrophic Concept has retrospectively been demonstrated through an increase in the relative number of articles per year discussing this subject. A following study of human metabolism has been shown to provide a further level of complexity through the comparison of methodological approaches, standing at odds with the frequent portrayal of BI as a constant topic of interest within Williams' career.

Free and Unequal: The Biological Basis of Individual Liberty has been shown to add a further element of depth to BI by focusing on the practical consequences arising from its lack of appreciation, rather than extensively attempting to prove its existence. Close reading and text-comparison has additionally revealed that, within this publication, Williams' overall tenor regarding BI changes. While previous publications on the subject largely contain some sort of appeal for the existence of BI and attempt to prove its relevance through data and concrete examples, this book portrays the concept as an uncontested fact. With a myriad of previous publications offering such proof discussed previously, impressing the importance of further studying and appreciating BI to science as well as general society has been revealed to be central within this publication through the in-depth analysis of its text. Within this context, *Free and Unequal* centrally contains the statement that the existence of differences must not be used to indicate superiority of one group of people over another – especially regarding "race" and sex. Historical literature on this time period has suggested that this is a highly progressive view for the average white, southern, middle-aged man of the early 1950s and may be drawn upon to indicate how BI supersedes societal prejudice and cultural norms in Williams' work. This facet of BI is central to *Free and Unequal*, a publication which essentially calls for the entire reorganisation of American society according to the ideals of BI. The issues of racism, gender equality, and political ideology have been shown to reappear in following publications as well as *Biochemical Individuality*, with this aspect of BI remaining relevant throughout the following years. While *Free and Unequal* does not extensively attempt to prove the existence of BI, *Biochemical Individuality* unifies the approaches of this and previous publications – first proving unequivocally that BI exists, and subsequently discussing its far-reaching consequences. This additionally provides evidence for the fact that *Biochemical Individuality* is a crystallisation point of Williams' research.

The pronounced influence of post-WWII McCarthyism on Williams' work and publications has additionally been specified above. With historical literature indicating how grants as well as government approval and support for scientific research were partly dependent on the executing scientists' political views at this time, Williams' views as expressed in his work on BI may have eased his path through the political landscape, though the exact extent of this effect is difficult to ascertain. Public statements regarding communism indicate a clear disdain for the forced unification imposed on the societies of the USSR from a scientific and moral standpoint. This is well within the common views on communism in McCarthyist America, though Williams additionally indicates the weaknesses of American society – something of a daring move at the time. His clear statements rebutting the principles of communism and its right to exist may have protected him from accusations of being communist himself, though this again is mere conjecture. These political statements are always made within the context of conceptual BI and aim to improve social equality and the functioning of society. Individuality, not politics, has clearly been shown to be Williams' central focus, with statements on topics outside of scientific research always only extrapolated from his research findings and field of expertise.

The analysis of personal correspondence has indicated how the dogmatic statements as contained in *Free and Unequal* are generally taken favourably by colleagues of various faculties, though the additional appreciation of peer reviews of this publication indicates that the extreme reorganisation as demanded therein isn't unanimously accepted and rather controversial. The influence of Williams' notoriety as an esteemed scientist becomes evident here as this book, containing much more extreme political statements which largely go against societal trends of the time, is most favourably received. Previously harshly criticised for much less extreme statements in *The Human Frontier*, Williams' headway with his research of BI and his renown in connection therewith may have affected how his work is later more widely accepted, though this cannot be proven unequivocally. Suggesting how an increased appreciation for BI may improve American society in general as well as the lives of all of its members, Williams' theories here have been indicated to explicitly stand at odds with the prevalent views of his time.

The growing importance of other scientists' work on individuality becomes clear in the publications of the final years before *Biochemical Individuality*'s publication discussed within this thesis. Close reading and diagram analysis have revealed the considerable extent to which the works of Barry Anson and Karl

Lashley are studied and discussed in *Biochemical Individuality*. This indicates how the research approach to BI can be described evidence-based in later years, appreciating the importance of outside research and statistical data, and using this to effectively bolster Williams' own scientific viewpoints. Previously having to largely rely on either historical or his own research, bibliographical analysis has shown how Williams' publications on BI begin to extensively draw on the work of other contemporary scientists, as well as his own, in the 1950s. This also indicates how the number of contemporaries sharing his viewpoints on individuality increases in the later stages of its development. Williams' own importance to his contemporaries additionally becomes evident when regarding his election as president of the American Chemical Society for 1957. Extensively travelling throughout the United States in order to be elected as such, personal communications have suggested how these journeys further increase his influence as well as spread his message of the importance of individuality within the scientific community.

Repeated analyses of the blood samples of 12 healthy individuals relate to Williams' ardent appeals for such research in previous decades. Historical accounts have been able to indicate the great progress of analytic technology, some of which have been shown to have been developed within the laboratories of DBUT. With primary ideas of individuality based on the observation of human phenomena, and an understanding thereof stated to be the ultimate purpose of BI, Williams "comes full circle" with this research. His scientific investigations progress not only in the complexity of the techniques they utilise, but also in the complexity of the organisms they study – reaching a final tier with "Normal Young Men" in 1955.

By comparing the contents of Williams' publications between 1919 and 1956 and *Biochemical Individuality: The Basis for the Genetotrophic Concept*, this analysis has been able to demonstrate how the latter unifies techniques and approaches of Williams' preceding research. Previous approaches such as "humanics" are disregarded here, and correspondence and peer reviews of the previous decades indicate the difficulties of fostering cooperation between the sciences. The title *Biochemical Individuality: The Basis for the Genetotrophic Concept* additionally indicates an aspect of realism: Williams presents his own knowledge and understanding of individuality fundamentally within the subject of biochemistry. Merely generalising within his proven field of expertise, this publication has not been found to present anything akin to a complete overview of human individuality regarding all fields following in-depth analysis of texts, diagrams, data, and bibliographies. Stressing the relevance of a holistic view, this analysis has merely ascertained that it reveals how the facts of biochemical variation have consequences for other scientific and academic fields while not, as discussed regarding prior publications, making overarching statements for these.

Expanding the volume of evidence for a wide variety of examples indicating interindividual human individuality, *Biochemical Individuality* builds upon the citations and examples presented in the preceding publications discussed above. Considering possible implications for biological, medical, dental, and psychiatric research, the analyses of bibliographies, texts, and reviews have shown that this work cements the relevance of Biochemical Individuality and the Genetotrophic Concept with irrevocable data. As evidenced by a plethora of positive reviews and amicable correspondences, *Biochemical Individuality* is a celebrated book to this day, with which Williams cements his reputation as an authority regarding individuality. Prior literature discussing his work has already largely concluded that *Biochemical Individuality* is also Roger Williams' most important publication on the topic of individuality – further indications for the truth of this statement have been made here.

Building on the research of Gadebusch Bondio (2017), and its delineation of Williams' central Genetotrophic Concept and his concept of Biochemical Individuality, the results of this study show an evolution of Williams' ideas throughout his career. This analysis indicates that Williams' concept of Biochemical Individuality undergoes a complex process of change; not singularly posed as a theory in 1956, but cumulatively developing between 1919 and 1956. This analysis has revealed previously undiscussed aspects of Williams' work: the inception and improvement of fundamental research methodologies and techniques in his study of BI. The appreciation of these can further indicate the pioneering nature of considerable proportions of Williams' work, as suggested in contemporary scientific publications (Badrick, 2021; Fitzgerald and Rountree, 2022; Giera et al., 2022; Patterson and Turnbaugh, 2014; Schloss, 2023). Through the knowledge gained from the analysis of Williams' private documents and recordings housed in the Dolph Briscoe Center for American History at The University of Texas, this study additionally provides an appreciation of Williams' subjective views on BI, and its development previously not studied, furthering the understanding of his thought- and research-processes.

Though biographical material on Roger Williams' life and work has previously been available and has therefore been able to provide central facts regarding Williams' life and work, the above investigation expands upon the existing knowledge through detailed analysis and discussion of Williams' research within the historical, political, cultural, and scientific timeframes. This allows for a more detailed understanding of the intrinsic and external influences on Williams' work, as well as a historical appreciation of his studies within the context of his direct predecessors and peers in the selected timeframe. Through their biographical memoir, Davis et al. (2008) have previously provided a framework overview of Williams' life and academic career.

The role and impact of individuals such as Warren Weaver, Robert R. Williams, Benjamin Clayton, and Linus Pauling is indicative of an external influence on Williams' theories, which are often described as entirely his own within the available biographical literature. The consequence of Williams' bodily limitations additionally provides a viewpoint of his distinctive working habits, while relativising the potential importance of outside influences. Building on this knowledge, this analysis has embedded Williams' research within the context of his (academic) peers, appreciating the role of other scientists and individuals within Williams' life and work. This presents Williams' work within the hitherto unexplored context of his own scientific process, discussing external influences alongside internal related developments. While the biographical information priorly accessible provided a narrative within which Williams' research can be discussed, this thesis has set out to provide more profound data pertaining to Williams' theories and publications through the in-depth discussion of specific publications and direct comparison thereof.

While similarities and differences to the work of other scientists have previously been denoted by Motulsky (2002) and Michl (2015), the influences of specific individuals, theories, and ideas have additionally been delineated here. The cited works respectively present Williams' work within the specific contexts of the developing pharmacogenomics and personalised medicine. This analysis has discussed Williams' work within its own context, therefore relating to subjects and individuals previously not studied. Though Michl (2015) has indicated the nature of Williams' research methodology and its ingenuity, the evolution of Williams' research has here been further illustrated using specific data and terminology from particular publications, denoting concrete starting and end points of multiple aspects of his research on BI. Similarly, this work expands upon Michl's (2015) discussion of the public ramifications of Williams' work through the detailed analysis of public reviews and the reception of his work in personal correspondence.

Multiple contemporary scientific publications portray *Biochemical Individuality* as the crystallisation point of Williams' research on BI (Badrick, 2021; Fitzgerald and Rountree, 2022; Giera et al., 2022; Patterson and Turnbaugh, 2014; Schloss, 2023); this has been conclusively exhibited within this work through the detailed examination of Williams' thought- and research-process alongside a comprehensive analysis regarding the contents of his publications between 1919 and 1956. A statistical analysis of the overarching subjects of

Williams' publications throughout his career additionally indicates the fluctuating relevance of BI within the aforementioned timeframe. This presents a contrast to the common depiction of Williams' work as involving two entirely separate fields of interest – fundamental biochemistry and Biochemical Individuality – and indicates the symbiotic nature of his varied topics of research.

The above results have revealed how the data presented within *Biochemical Individuality* represents information on variation collected by Williams across previous decades, discussing the ways in which new material published primarily therein is additionally relevant. To this end, corresponding references within his works have been presented in tabular form.[354] These tables are visually indicative of the pronounced relevance of Williams' own research to his understanding of BI, providing evidence for the accumulative nature of BI. A bar graph depicting the contents of Williams' publications throughout his career indicates the growing relative importance of BI in Williams' research. Alongside the analysis of the texts and bibliographies of Williams' works between 1919 and 1956, primary sources housed at the Dolph Briscoe Center for American History at The University of Texas and pertinent to the timeframe of this research have been viewed, recorded, and analysed according to the scope and developed research questions for this study. Various aspects of the developing concept have additionally been related to the book most often described as Williams' decisive publication on BI in historical and contemporary literature: *Biochemical Individuality: The Basis for the Genetotrophic Concept*. Though Williams' work has heretofore been contextualised and classified historically regarding the scientific aspects of his concepts by Gadebusch Bondio and Spöring (2017), the political influences and ramifications of Williams' theories are newly appreciated here.

The presented evidence, collected by close reading of publications and the analysis of auxiliary material, has confirmed the hypothesis that BI was evolved and refined as a concept throughout the career of Roger J. Williams, finally culminating in the publication of *Biochemical Individuality: The Basis for the Genetotrophic Concept* in 1956. Important alterations to and developments of Williams' theories have been explored and analysed by relating the language and substance of his earlier works to the resulting publication on individuality, the significance of these variations for his final concept of Biochemical Individuality having been stressed through short abstracts concluding each

---

354 See Tab. 2: Articles Cited by *Biochemical Individuality* Appearing in Prior Publications.

chapter. Furthermore, contextualisation has been provided wherever pertinent, discussing Williams' work within the timeframe of its inception. Examining the pronounced influence of the scientific zeitgeist on his research, this work has indicated the modulating effects of Williams' scientific and academic contemporaries. The weight of political and cultural stimuli in the United States has additionally been deliberated through the identification and analysis of Williams' own political statements considered in relation to corresponding landscapes of the time. The evaluation of the bibliographical information provided in Williams' publications has shown a significant overlap between this definitive work and his previous research. All of Williams' own publications cited in *Biochemical Individuality* have been subject to in-depth discussion, with modifications and parallels between the publications denoted accordingly.

Though the research for and writing of the preceding investigation has been executed with utmost care, certain methodological and circumstantial limitations remain. Regarding circumstantial limitations, the travel ban issued due to the COVID-19 pandemic during the initial research phase delayed the sighting of the original material in the Dolph Briscoe Center for American History at The University of Texas. A planned research trip was postponed on multiple occasions, causing further interruptions in the study of this works' hypotheses before seeing the vital material in Austin. Large portions of this work were therefore initially produced according to assumptions made on the basis of Williams' published manuscripts, the truthfulness of which could only later be assessed by the appreciation of the primary sources. Therefore, certain sections had to be re-written and changed because of new or alternative insights. Similarly, though efforts were made to acquire first editions of all of Williams' works, the age and rarity of some of his textbooks meant that, in some cases, only later editions were available.

Concerning methodological limitations, the scope of this dissertation primarily is limited to documents published until 1956, with the analysis of Williams' later works potentially offering further insight. This work has been able to provide an overview of Williams' career between 1919 and 1956, while the timeframe between 1956 and 1988 has largely not undergone consideration. Though all of Williams' published books have been considered as to their merit regarding the origins and development of Biochemical Individuality as a concept, this analysis has made no attempt to discuss their contents as they pertain to the further progression of BI, or even whether Williams' concept evolves further at all. The same pertains to all boxes at the Briscoe Center solely containing documents from the years following 1957; these were not viewed and analysed, meaning that some evidence may have remained unsighted. This work can therefore not attempt to make universal claims on Williams' entire career, though it may provide a basis on which such broader work can build. Similarly, this methodology limits the validity and significance of statements regarding the third research question of this thesis, as not all works following the publication of *Biochemical Individuality* have been considered. A work discussing Williams' entire legacy could provide a more profound view, considering his publications and theories, as well as the importance and station of *Biochemical Individuality* within the context of his entire career, which would allow for further reaching statements regarding BI.

As the product of all previous research, the publication of *Biochemical Individuality* signifies a metaphoric crossroads in Williams' career. While his research on nutrition, alcoholism, individuality, and other topics of personal interest continues after 1956, this is based on the facts and assumptions presented within the book's two-hundred and nine pages. Though the progression of Williams' later investigations and career has great potential for further studies and academic work, the scope of this thesis does not allow for analysis past the publication of *Biochemical Individuality*. Remaining academically active to the end of his life in 1988, Williams publishes eleven books and one-hundred and twenty-one articles following *Biochemical Individuality*, with nutrition and education – in accordance with his concept of individuality – at the forefront of his studies (Davis, 2003c, 2003d, 2003e). Analogously, the Dolph Briscoe Center for American History at The University of Texas at Austin houses a further number of unsighted boxes containing documents from the decades following the publication of *Biochemical Individuality*.

Only partially relevant to the central theme of this thesis, Williams' studies on pantothenic acid and the B vitamins have not been analysed for their biochemical and scientific merit. The Roger Williams Papers contain a large number of scientific research notes and logs for this work, as well as a wide array of correspondence regarding these studies. A most influential discovery within the field of biochemistry, this research could be reviewed in further detail from a biochemical, medical, ecotrophological, and historical viewpoint. Similarly, Williams' influence on the developing field of nutrition in the 20th century has also been omitted from this thesis. From fundamental research on vitamin biochemistry and books on nutrition, to his later contributions as part of panels and boards of nutrition, such study could provide further understanding and improve the appreciation for this line of his work. Furthermore, an exploration of Williams' position within the framework of the development of evidence-based medicine and epigenetics could offer an interesting avenue for research. Williams repeatedly draws attention to the fact that concrete evidence of human individuality is severely lacking, and that human beings are therefore regularly mistreated and misunderstood. He emphasises the importance of evidence to medicine and medical treatment on multiple occasions, presenting ideas akin to the principles upon which medical guidelines are developed today – the study of his potential role within the development of EBM may prove consequential.

Overall, it is difficult to gauge the extent of the influence of Roger Williams on contemporary ideals of human individuality. Though the role of human geneticists is often highlighted in works on the history of personalised medicine (Emmert-Streib, 2013; Jain, 2009; Jones, 2013a; Marshall, 1997; Motulsky, 2002; Visvikis-Siest et al., 2020), the concept of tailoring medical treatments to the specific needs of biochemically unique individuals must base itself on the knowledge that these biochemical individualities exist. Similarly, the development of educational programs with increased attention to the individuality of students reflect the principles of Williams' publications (Department of Education and Training Victoria, 2023; U.S. Department of Labor, 2023). His concepts must therefore provide a basis for modern opinions on individuality, with his scientific influences regularly appreciated in the 21st century.[355] As a prominent writer of non-fiction for laypeople in the 20th century (Fowler, 1988), Williams' works contributed to the understanding of individuality of previous generations, and therefore also influences the perceptions of more recent times. Recognised as an effectively concise and well-structured summation of human individuality, *Biochemical Individuality* was reissued in 1998, when individualised/personalised medicine had established itself as a promising new philosophy (Williams, 1998).

Reminiscing on more than 70 years of work and research, Williams describes how much joy he has derived from his work on BI, especially all attempts to "sell" the concept to the scientific community and public (Hodge and Williams, 1980a). Feeling the importance of BI had not made a large enough impression on the scientific community, Williams ultimately describes his efforts as a failure. This summation of the "uphill battle" of drumming up real scientific and societal interest in his concepts seems a rather bleak conclusion when his work is becoming more important today than ever before (Hodge and Williams, 1980a). His successful application of the Genetotrophic Concept is indicative therefore, even if Williams' view of alcoholism as a purely genetotrophic disease is not shared by contemporary medical guidelines (World Health Organization, 2018, p. 9). While the role of genetics in an individual's propensity towards alcoholism has been established today, the aetiology of alcoholism is widely regarded to be multifactorial (Tawa et al., 2016). Though Williams' proposed treatment of

---

355 See Section 1.1. Why Individuality?.

alcoholism through vitamin supplementation is no longer seen as viable today, the administration of thiamine (vitamin B_1), and vitamin supplementation generally, plays an important role in the prophylaxis of ethanol-induced cerebellar ataxia (Mitoma et al., 2021, pp. 5–6). Aspects of Williams' concepts are therefore still reflected in the prophylaxis of alcohol-related sequelae today. Similarly, Williams' strides in personalised nutritional optimisation, later central to his continuing research on BI, remain an important pillar of nutritional theory and practice today (Bland, 2019a, 2019b; Gasta, 2020; Johnson and Hand, 2020; de Las Hazas and Dávalos, 2022; Schloss, 2023; Steg et al., 2022). When considering the fact that western societies are, by and large, coming to realise the importance of treating individuals as such, Williams' pessimistic view must then not be taken more than 40 years later, and may become even less true for future generations. With the increasing influence of "Gen Z" and individuality at the forefront of this generation's priorities,[356] the importance of Williams' scientific and cultural contribution, and the understanding thereof, becomes ever more significant.

---

356   See Chapter 1. Introduction.

15 Summary

This thesis reconstructs and discusses the research of Roger J. Williams, a biochemist and professor at The University of Texas at Austin, and the process of development behind his concept of Biochemical Individuality. The fundamental theories on which this conjugate term is based summarise all forms of variation shown by human beings on a fundamental metabolic level and explore the consequences of these deviations for science and society. Roger J. Williams' theories on Biochemical Individuality are elemental to our contemporary concepts of human distinctiveness, forming a basis for important aspects of what we call "personalised medicine" today. As a pioneer within the budding scientific field of biochemistry, Williams explored, collected, and analysed the extent to which human beings differ on a (bio)chemical level, while additionally investigating the effect of this individuality on science, politics, and society.

Prior works discussing Williams and his research of individuality focus on his theories in their finalised form as set out by his well-known book *Biochemical Individuality: The Basis for the Genetotrophic Concept*, while this analysis in turn reconstructs the process of development behind these ideas from inception to conclusion. Furthermore, this work deliberates whether the book *Biochemical Individuality* signifies a crystallisation point within the overall context of Williams' study of individuality. Chronologically examining all of Williams' publications and personal documents between the years 1919 and 1956 and discussing these within their historical, political, cultural, and scientific contexts, this analysis investigates the terminology, bibliographical data, and scientific material of these documents in order to illustrate how Williams' theories and ideas morph over time, while appreciating internal and external influences on this process.

This research has shown that the concept of Biochemical Individuality undergoes a complex process of development throughout Williams' career, ultimately culminating in the publication of *Biochemical Individuality: The Basis for the Genetotrophic Concept* in 1956. The earliest concepts of Biochemical Individuality are entirely devoid of structure and purely based on personal experiences and intuitions. Alongside his primary studies of the fundamental biochemistry of vitamins and their functions from 1919 onwards, first elements of structure begin to appear within a one-off publication discussing individuality in taste perceptions published in 1931. With individuality more of a niche matter of interest to Williams, there is only little evidence of concrete progress in this matter until more than a decade later. Evidence has shown, however, that Williams' early biochemical research contributes to his understanding of variation in living organisms and provides him with expertise later relevant to his further studies. In 1946, Williams publishes a widely critiqued book entitled *The Human Frontier: A New Pathway for Science Toward a Better Understanding of Ourselves*, in which he not only presents large quantities of evidence for human variation, but also discusses how the more profound appreciation of these differences is of absolute essence to human societies. Here, the influence of specific scientific theories and outside individuals becomes clear, and Williams begins to structure his thoughts into a preliminary concept. Williams here draws attention to the lack of active research on human biochemical individuality, this appeal for more research remains a central feature of all of his following publications.

From this point onward, variation becomes Williams' research focus, with the majority of his publications discussing Biochemical Individuality (which appears as a conjugate term in 1947 for the first time) in some form. Beginning to study the effect of interindividual variation on propensities toward alcoholism in rats, Williams benefits from his early biochemical training and begins to make a name for himself as an expert on human variation. The political climate of the post-WWII United States has been shown to lead to further diversifying in the reach of his concepts. The Genetotrophic Concept is first posed in 1949 and becomes a central subject of study for the rest of Williams' career. Suggesting that diseases may stem from individual nutritional deficiencies, Williams again profits from his early nutritional research and begins to study the medical potentialities of his theories. Publishing what can essentially by described as a political manifesto in a book entitled *Free and Unequal: The Biological Basis of Individual Liberty* in 1953, Williams' concepts become further dominant and profound. Here, he discusses the overwhelming evidence against exclusion or discrimination on the basis of sex or "race" and societal shortcomings while acting within the political context of the Cold War. The final years before *Biochemical Individuality*'s publication are shaped by Williams' bid to become president of the American Chemical Society and his own studies on variation in human males.

Biochemical Individuality: The Basis for the Genetotrophic Concept is published in 1956, presenting Williams' finalised concepts and discussing their consequences for a variety of scientific fields and general society. The analysis of Williams' publications has shown that this book represents an abridged summary of all previous research, reflecting aspects of all stages of Biochemical Individuality's development. It is therefore a clear crystallisation point of his research on human variation and is in accordance with the previously available academic literature on the subject. This work has also indicated how his theories progress and develop, expanding on the previously purported depictions of Biochemical Individuality as a finished product by the in-depth discussion and depiction of Williams' research process alongside the relativisation of the influence of external theories, circumstances, and individuals. These results must be seen in the light of circumstantial and methodological limitations, namely the retrospective sighting of archive material due to pandemic travel restrictions, the lack of appreciation of publications outside of the predefined timeframe, and the unavailability of certain historical documents. Williams' research from 1956 onwards up to his death in 1988 have not been studied here and offer further potential for academic deliberation. Williams' role in the development of evidence-based medical practices could offer further insights, as well as the relevance of his vitamin studies from the viewpoint of multiple alternate specialities. Williams' studies and concepts of individuality largely remain applicable to medical and cultural conceptions of individuality to this day and can provide a historical perspective on central philosophies for current and future generations.

16 Annex

16.1 Interview with Donald R. Davis

The following pages are based upon the recording and notes made during an interview with Donald R. Davis via telephone on the 28th of January 2023. Where not pertinent to this thesis, passages of conversation have been omitted, while minor grammatical changes have been made to improve comprehensibility.

Brand: I'd like to start out by thanking you for all of your help with my dissertation over the past years, it has really been invaluable, and I am sure that I would not have come nearly as far without your assistance.

Davis: You're very welcome, I was very glad to help. I appreciate your interest and look forward to your dissertation. I'm really pleased to help and have been delighted by the work you've been doing.

Brand: We have spoken about your work at UT a few times via E-Mail, but could you remind me of what brought you to Austin to work with Roger Williams?

Davis: I grew up in the desert of California just north of Los Angeles, and most of my family is still in that area. The reason I'm in Texas is because Roger Williams worked here. Before that I was teaching at the University of California. I first came to Austin in 1973 on a sabbatical leave working for Williams. He invited me to come back, and so I did in 1974. I stayed on at the University until 2007.

Brand: So, Williams was still active when you arrived?

Davis: When I arrived in 1974, Williams was, surprisingly, still quite active. He was about 80 at that point I think, but he still had an active research group and was working on several books. He didn't quit easily.

Brand: I was astounded to see that his last publication was written in the final year of his life. He really must have had a phenomenal work ethic and drive.

Davis: Yes, it was unbelievable. He was very single-minded and had a lot that he wanted to tell the world.

Brand: What would you say was the central message he wanted to convey to the world?

Davis: I think it was individuality that was most on his mind. The most recent manuscript that was never published was on individuality. He was best known scientifically for the discovery and naming of pantothenic acid, but I think if you look at the titles of his books, his career was more focused on all aspects of individual differences.

Brand: It's interesting you say that, because that is the same conclusion I have come to during my research. Most of the articles in which he is in some way mentioned largely discuss his work on pantothenic acid, even though this was really just the beginning of his career. Did Williams speak about individuality with you and other colleagues often? Or was it "just" something he thought and wrote about.

Davis: As far as I know, none of his colleagues were very much focused on individuality. I know that he did share manuscripts with many colleagues, asking them to review and comment on them. But I don't think he had any "soulmates" so to speak, to talk about individuality. He had a long-term office assistant, Margaret Biesele; she probably was a sounding board for him. In later years, I was too. He would show me things he was working on and would ask for comments.

Brand: I did find a great number of letters from people who he had sent manuscripts to, some including very famous names in the archive. Linus Pauling, Bertrand Russel, and Lyndon B. Johnson to name a few.

Davis: He did send them out to a lot of people. He spent a lot of time on his manuscripts. He was always interested in people's reactions and what changes he might make inspired by them.

Brand: How important were these comments from others to him? What value were they attributed?

Davis: They were of great value, and they were usually integrated into later versions, though not necessarily always. He was always looking for something that he thought was valuable. Near the end of his life, he was in a nursing home and still working on a manuscript. He still had it all in his head and would sometimes ask us to come and help him revise a certain part of it that he was working on in his mind. That's one of the really valuable things about his writing – he didn't just dash it off. He thought about it a lot and got feedback. I think that improved the longevity of the things that he wrote.

Brand: I did find it remarkable how well-written all of his books are, they really are quite eloquent. Especially coming from someone with a scientific background that was an aspect that really stood out.

Davis: (Laughs), you know as a young man he was interested in the ministry. He had a certain preacher aspect to him. His parents were missionaries, so this always was interesting to him.

Brand: I had a very similar feeling when watching his interview tapes in the archives.

Davis: I'm glad you got a chance to see those.

Brand: Coming back to another point you raised earlier, you said that his colleagues at the biochemical institute weren't as concerned with individuality. Did Williams give you and your colleagues the freedom to essentially study whatever you were interested in?

Davis: He was the founding director of the biochemical institute and as I understand it, he always selected his co-workers based upon their outstanding scientific interest and abilities. So, the emphasis was on science and not on individuality there. He pretty much pioneered this aspect of individuality. There will not have been very many people he could have invited to work for him and collaborate with, because he was really leading the way.

Brand: That is reflected by most accounts of the first developments of personalised medicine. He sticks out as a biochemist, in a relatively homogenous group of human geneticists. How did you first come across Roger Williams and his work?

Davis: I first learned about him, because, though my training was originally in physical chemistry and I was teaching chemistry at the University of California, Irvine, my interests were beginning to change more toward human development and one of my students brought me a book of his. This triggered my interest, and so I quickly came across his "Nutrition Against Disease", in which he writes a lot about individuality. And so, I wrote to him asking if I could work at his institute on a sabbatical leave in 1973.

Brand: What kind of work did you do while you were there on your sabbatical?

Davis: I was only there for about 3 months and did an experiment feeding rats an average American diet vs. multiple supplemented diets. We later published an article in the *American Journal of Clinical Nutrition* about the results. He gave me a lot of freedom to work on whatever I wanted, even for just the short sabbatical leave.

Brand: Leaving others the freedom to research and work on whatever they'd like seems really a central to his leadership style.

Davis: I think he gave a lot of freedom to everyone he invited to work for him, leaving them to follow whatever their interests were.

Brand: That fits very well into the whole context of individuality.

Davis: Yes, he let them work according to their own individualities. *Brand*: An aspect I found most interesting when first starting my research, was on the list of publications on your website. Williams was essentially published in every meaningful medical and scientific journal, but also in publications like "Vogue". *Davis*: Yes, he also published in Saturday Review and Texas Monthly, so he wrote for lay-people as well as scientists.

Brand: Did he ever speak about how important writing for a public audience was to him?

Davis: I don't recall him specifically talking about that, but I think it was very important to him by implication. *Nutrition Against Disease*, *You Are Extraordinary*, *The Human Frontier*, and *Free and Unequal* were definitely aimed at a lay audience rather than a professional audience. *Biochemical Individuality* itself was rather scientifically oriented.

Brand: Speaking about *Biochemical Individuality*, that book always appeared to me as being a real crystallisation point in which all of his research of the previous decades came together.

Davis: Yes, I think you're right. He usually spoke about individuality in a relatively narrow context. But I know that he visited a lot of people while writing *Biochemical Individuality* in order to learn about the particular areas they were interested in. It wasn't something you could go into a library for and look up, you really had to go and talk to these people. I don't know how long he spent on it, but I know he spent a lot of time travelling around and meeting with people. For example, he visited Barry Anson, who's work on anatomical variation inspired parts of *Biochemical Individuality*. I have many of those papers myself.

Brand: One thing I always found interesting in Williams' work, was that he never feared to deviate from what was regarded as "normal" procedure in his work. Something you have written about, and Williams himself mentions, is his bad eyesight. Could you expand on how this may have influenced his innovations?

Davis: Yes, I remember him saying he liked to think about things rather than read about them. Some people spent a lot of time researching, he spent a lot of time thinking instead of searching for something that someone else had written. He was more interested in developing his own ideas.

Brand: I guess one might say that his own individuality benefited his work on individuality.

Davis: Definitely. When I arrived in Austin, his eyesight had further diminished in addition to his aniseikonia. He had developed macular degeneration, so that he could only barely read. For work, he used a magnifying device to help him read anything. He had very little vision left by 1973. It didn't stop him walking home for 2 miles every day, because he still had some peripheral vision. He liked to get exercise that way.

Brand: One final point I'd like to ask you about, is the effect of *Biochemical Individuality* on Williams' later work. My dissertation finishes with its publication in 1956, how important do you feel that the book was for the work that followed?

Davis: All books that appeared afterward were certainly influenced by *Biochemical Individuality*. As he had been developing his ideas for quite a while, *Biochemical Individuality* really brought these all together in a scientific sense. This knowledge then formed a basis for his later work. Individuality was still his most important topic later in his career. When someone in our institute would make a statement not entirely cognitive of the facts of individuality, he would gently remind us of how important it was. It was always on his mind.

Brand: Thank you so much for taking the time to talk to me, I will make sure to send you a copy of my dissertation when it's handed in!

Davis: I am very happy to do that, Georg. Again, I appreciate what you've done and am glad that your trip to Austin has been a success.

17 List of figures

Fig. 1: Letter by Warren Weaver to Williams on 28.01.1946 critiquing Williams' use of the term "humanology" in an early version of The Human Frontier, first page; Dolph Briscoe Center for American History, The University of Texas at Austin, Roger Williams Papers, Box 88-087/26a, Folder: Correspondence Concerning Humanics Sept. 1945 – March 1946 34

Fig. 2: Letter by Warren Weaver to Williams on 28.01.1946 critiquing Williams' use of the term "humanology" in an early version of *The Human Frontier*, second page; Dolph Briscoe Center for American History, The University of Texas at Austin, Roger Williams Papers, Box 88-087/26a, Folder: Correspondence Concerning Humanics Sept. 1945 – March 1946 35

Fig. 3: Statement of Sales by University of Oklahoma Press on 01.07.1950 indicating that What To Do About Vitamins had sold a total of 2735 copies; Dolph Briscoe Center for American History, The University of Texas at Austin, Roger Williams Papers, Box 88-087/4 67

Fig. 4: Statement of Sales by University of Oklahoma Press on 01.07.1953 indicating that *What To Do About Vitamins* had sold a total of 3007 copies; Dolph Briscoe Center for American History, The University of Texas at Austin, Roger Williams Papers, Box 88-087/4 67

Fig. 5: Title Cover of the 1946 Hardback Version of The Human Frontier: A New Pathway for Science Toward a Better Understanding of Ourselves (Williams, 1946a) 76

Fig. 6: Title Cover of the 3rd Edition of *An Introduction to Organic Chemistry* (Williams, 1935a) 77

Fig. 7: Telegram by Paul De Kruif to Williams on 14.12.1945 confirming Harcourt, Brace and Co. is interested in publishing *The Human Frontier*; Dolph Briscoe Center for American History, The University of Texas at Austin, Roger Williams Papers, Box 88-087/26a, Folder: Correspondence Concerning Humanics Sept. 1945 – March 1946 79

Fig. 8: Letter by John D. Newsem to Paul De Kruif on 09.01.1946 confirming that Harcourt, Brace and Co. is interested in publishing *The Human Frontier*; Dolph Briscoe Center for

	American History, The University of Texas at Austin, Roger Williams Papers, Box 88-087/26a, Folder: Correspondence Concerning Humanics Sept. 1945 – March 1946	80
Fig. 9:	Letter by Williams to D. M. McKeithan on 20.05.1946 asking for views on a manuscript of *The Human Frontier*; Dolph Briscoe Center for American History, The University of Texas at Austin, Roger Williams Papers, Box 88-087/26a, Folder: Humanics Article Faculty Reactions	82
Fig. 10:	Letter by John Foster Dulles to Williams on 24.10.1946 indicating his interest in *The Human Frontier*; Dolph Briscoe Center for American History, The University of Texas at Austin, Roger Williams Papers, Box 88-087/26a, Folder: Famous Names (Humanics)	106
Fig. 11:	Letter by John D. Rockefeller III to Williams on 24.04.1947 thanking him for sending a copy of *The Human Frontier* and indicating a meeting two weeks prior; Dolph Briscoe Center for American History, The University of Texas at Austin, Roger Williams Papers, Box 88-087/26a, Folder: Famous Names (Humanics)	107
Fig. 12:	Letter by Lyndon B. Johnson to Williams on 16.06.1950 indicating his interest in Williams' address to the Philosophical Society of Texas; Dolph Briscoe Center for American History, The University of Texas at Austin, Roger Williams Papers, Box 88-087/41, Folder: LETTERS: RE "SHALL WE PIONEER TOO?"	143
Fig. 13:	Publications by Roger J. Williams (1919–1956)	150
Fig. 14:	Letter by Lorene Lane Rogers to T. J. Boman on 09.06.1952 indicating Williams' absence following Hazel's death; Dolph Briscoe Center for American History, The University of Texas at Austin, Roger Williams Papers, Box 88-087/23a, Folder: Alcoholism – Inquiries, Reports, Correspondence, Physicians, Scientists, Governmental Agencies, 1955 –	165
Fig. 15:	Letter by Williams to Frank H. Wardlaw on 25.01.1952 including early ideas for the title of *Free and Unequal*, page one; Dolph Briscoe Center for American History, The University of Texas at Austin, Roger Williams Papers, Box 88-087/1k, Folder: The University of Oklahoma Press (Lottinville) 1950–1951	175
Fig. 16:	Letter by Williams to Frank H. Wardlaw on 25.01.1952 including early ideas for the title of *Free and Unequal*, page	

	two; Dolph Briscoe Center for American History, The University of Texas at Austin, Roger Williams Papers, Box 88-087/1k, Folder: The University of Oklahoma Press (Lottinville) 1950–1951	176
Fig. 17:	Letter by Williams to Frank H. Wardlaw on 25.01.1952 including early ideas for the title of *Free and Unequal*, page three; Dolph Briscoe Center for American History, The University of Texas at Austin, Roger Williams Papers, Box 88-087/1k, Folder: The University of Oklahoma Press (Lottinville) 1950–1951	177
Fig. 18:	Letter by Williams to Frank H. Wardlaw on 25.01.1952 including early ideas for the title of *Free and Unequal*, page four; Dolph Briscoe Center for American History, The University of Texas at Austin, Roger Williams Papers, Box 88-087/1k, Folder: The University of Oklahoma Press (Lottinville) 1950–1951	178
Fig. 19:	Carbon Copy of a Letter by Julian Huxley to Williams on 21.01.1957 indicating his interest and appreciation for the contents of *Biochemical Individuality*; Dolph Briscoe Center for American History, The University of Texas at Austin, Roger Williams Papers, Box 88-087/26a, Folder: Famous Names (Humanics)	212

18 List of tables

Tab. 1: Comparison Between Overlapping Citations in *The Human Frontier* and *Biochemical Individuality* 86

Tab. 2: Articles Cited by *Biochemical Individuality* Appearing in Prior Publications 112

Tab. 3: Overlapping References from *Biochemical Individuality* and Publications in 1956 196

Tab. 4: Citations of Williams' Own Publications in *Biochemical Individuality* 205

19 References

19.1 Unpublished Materials

Barnard CI. Letter to Williams RJ, 14 June 1950. Roger Williams Papers, Box 88-087/41, Folder: LETTERS: RE "SHALL WE PIONEER TOO?", Dolph Briscoe Center for American History, The University of Texas at Austin.

Baruch BM. Letter to Williams RJ, 24 October 1946. Roger Williams Papers, Box 88-087/26a, Folder: Famous Names (Humanics), Dolph Briscoe Center for American History.

Beerstecher Jr. E. Letter to Williams RJ, 10 May 1950. Roger Williams Papers, Box 88-087/47b, Folder: Williams Personell: Miscellaneous, Dolph Briscoe Center for American History, The University of Texas at Austin.

Benner FV. Letter to Williams RJ, 30 August 1955. Roger Williams Papers, Box 88-087/47b, Folder: Williams ACS Tour 1955, Dolph Briscoe Center for American History, The University of Texas at Austin.

Bogert MT. Letter to Williams RJ, 14 June 1950. Roger Williams Papers, Box 88-087/41, Folder: LETTERS: RE "SHALL WE PIONEER TOO?", Dolph Briscoe Center for American History, The University of Texas at Austin.

Books Received. N Engl J Med 1945. Roger Williams Papers, Box 88-087/1k, "What To Do About Vitamins", Dolph Briscoe Center for American History, The University of Texas at Austin.

Camp WG. Letter to Williams RJ, 27 October 1947. Roger Williams Papers, Box 88-087/26a, Folder: Comments from Important Men Regarding The Human Frontier, Dolph Briscoe Center for American History, The University of Texas at Austin.

Clark TC. Letter to Williams RJ, 28 October 1946. Roger Williams Papers, Box 88-087/26a, Folder: Famous Names (Humanics), Dolph Briscoe Center for American History, The University of Texas at Austin.

Connally TT. Letter to Williams RJ, 23 June 1950. Roger Williams Papers, Box 88-087/41, Folder: LETTERS: RE "SHALL WE PIONEER TOO?", Dolph Briscoe Center for American History, The University of Texas at Austin.

Cook RC. Letter to Williams RJ, 9 January 1943. Roger Williams Papers, Box 88-087/1l, Folder: Williams, Roger J. – Personal Correspondence, etc., Dolph Briscoe Center for American History, The University of Texas at Austin.

Cullum LH. Letter to Williams RJ, 8 June 1950. Roger Williams Papers, Box 88-087/41, Folder: LETTERS: RE 'SHALL WE PIONEER TOO?', Dolph Briscoe Center for American History, The University of Texas at Austin.

Dickson J. Letter to Williams RJ, 7 February 1944. Roger Williams Papers, Box 88-087/1l, Folder: Williams, Roger J. – Personal Correspondence, etc., Dolph Briscoe Center for American History, The University of Texas at Austin.

Dr. Williams, '57, President of ACS. Houston Press 12 December 1955: Roger Williams Papers, Box 88-087/47b, Folder: Clippings, Dolph Briscoe Center for American History, The University of Texas at Austin.

Dulles JF. Letter to Williams RJ, 24 October 1946. Roger Williams Papers, Box 88-087/26a, Folder: Famous Names (Humanics), Dolph Briscoe Center for American History, The University of Texas at Austin.

Faller H. Letter to Williams RJ, Undated. Roger Williams Papers, Box 88-087/1l, Folder: Williams, Roger J. – Correspondence, Dolph Briscoe Center for American History, The University of Texas at Austin.

Fifield Jr. JW. Letter to Williams RJ, 14 June 1950. Roger Williams Papers, Box 88-087/41, Folder: LETTERS: RE "SHALL WE PIONEER TOO?", Dolph Briscoe Center for American History, The University of Texas at Austin.

Frederick JT. I've Been Reading. Sun 17 June 1945. Roger Williams Papers, Box 88-087/1k, "What To Do About Vitamins", Dolph Briscoe Center for American History, The University of Texas at Austin.

Fresneda MV. Alcoholismo: Nutrition and Alcoholism, por R. J. Williams. Guía Ter Cuba 1953: 265. Alcoholism – Inquiries, Reports, Correspondence, Physicians, Scientists, Governmental Agencies, 1955 –, Dolph Briscoe Center for American History, The University of Texas at Austin.

Fulbright JW. Letter to Williams RJ, 13 June 1950. Roger Williams Papers, Box 88-087/41, Folder: LETTERS: RE "SHALL WE PIONEER TOO?", Dolph Briscoe Center for American History, The University of Texas at Austin.

Giannini AP. Letter to Williams RJ, 25 October 1946. Roger Williams Papers, Box 88-087/26a, Folder: Famous Names (Humanics), Dolph Briscoe Center for American History, The University of Texas at Austin.

Good Things To Know About Vitamins. Book Mon 1945. Roger Williams Papers, Box 88-087/1k, "What To Do About Vitamins", Dolph Briscoe Center for American History, The University of Texas at Austin.

Green W. Letter to Williams RJ, 16 January 1947. Roger Williams Papers, Box 88-087/26a, Folder: Famous Names (Humanics), Dolph Briscoe Center for American History, The University of Texas at Austin.

Hart JP. Letter to Williams RJ, 6 December 1950. Roger Williams Papers, Box 88-087/41, Folder: LETTERS: RE "SHALL WE PIONEER TOO?", Dolph Briscoe Center for American History, The University of Texas at Austin.

Hodge B, Williams RJ. Conversations with Roger Williams Biography (Part One) 1980a. Roger Williams Papers Box 88-087/54, Dolph Briscoe Center for American History, The University of Texas at Austin.

Hodge B, Williams RJ. Conversations with Roger Williams Biography (Part Two) 1980b. Roger Williams Papers Box 88-087/54, Dolph Briscoe Center for American History, The University of Texas at Austin.

Hodge B, Williams RJ. Conversations with Roger Williams Tape #2: Individuality 1980c. Roger Williams Papers Box 88-087/54, Dolph Briscoe Center for American History, The University of Texas at Austin.

Hodge B, Williams RJ. Conversations with Roger Williams Tape #5: Alcoholism and Nutrition 1980d. Roger Williams Papers Box 88-087/54, Dolph Briscoe Center for American History, The University of Texas at Austin.

Holmes JH. Letter to Williams RJ, 24 October 1946. Roger Williams Papers, Box 88-087/26a, Folder: Famous Names (Humanics), Dolph Briscoe Center for American History, The University of Texas at Austin.

Hoover JE. Letter to Williams RJ, 30 October 1946. Roger Williams Papers, Box 88-087/26a, Folder: Famous Names (Humanics), Dolph Briscoe Center for American History, The University of Texas at Austin.

Horton D. Letter to Williams RJ, 14 June 1950. Roger Williams Papers, Box 88-087/41, Folder: LETTERS: RE "SHALL WE PIONEER TOO?", Dolph Briscoe Center for American History.

Hutchins RM. Letter to Williams RJ, 10 July 1950. Roger Williams Papers, Box 88-087/41, Folder: LETTERS: RE "SHALL WE PIONEER TOO?", Dolph Briscoe Center for American History, The University of Texas at Austin.

Huxley J. Letter to Williams RJ, 21 January 1957. Roger Williams Papers, Box 88-087/26a, Folder: Famous Names (Humanics), Dolph Briscoe Center for American History, The University of Texas at Austin.

Jester BH. Letter to Williams RJ, 18 November 1946. Roger Williams Papers, Box 88-087/26a, Folder: Famous Names (Humanics), Dolph Briscoe Center for American History, The University of Texas at Austin.

Johnson LB. Letter to Williams RJ, 16 June 1950. Roger Williams Papers, Box 88-087/41, Folder: LETTERS: RE "SHALL WE PIONEER TOO?", Dolph Briscoe Center for American History, The University of Texas at Austin.

Kettering CF. Letter to Williams RJ, 28 October 1946. Roger Williams Papers, Box 88-087/26a, Folder: Famous Names (Humanics), Dolph Briscoe Center for American History, The University of Texas at Austin.

de Kruif P. Letter to Williams RJ, 8 November 1945a. Roger Williams Papers, Box 88-087/26a, Folder: Correspondence Concerning Humanics Sept. 1945 – March 1946, Dolph Briscoe Center for American History, The University of Texas at Austin.

de Kruif P. Letter to Williams RJ, 5 December 1945b. Roger Williams Papers, Box 88-087/26a, Folder: Correspondence Concerning Humanics Sept.

1945 – March 1946, Dolph Briscoe Center for American History, The University of Texas at Austin.

de Kruif P. Letter to Williams RJ, 14 December 1945c. Roger Williams Papers, Box 88-087/26a, Folder: Correspondence Concerning Humanics Sept. 1945 – March 1946, Dolph Briscoe Center for American History, The University of Texas at Austin.

McKeen Cattell J. Letter to Williams RJ, 17 May 1943. Roger Williams Papers, Box 88-087/1l, Folder: Williams, Roger J. – Personal Correspondence, etc., Dolph Briscoe Center for American History, The University of Texas at Austin.

Newsem JD. Letter to de Kruif P, 9 January 1946. Roger Williams Papers, Box 88-087/26a, Folder: Correspondence Concerning Humanics Sept. 1945 – March 1946, Dolph Briscoe Center for American History, The University of Texas at Austin.

Nichols CA. Letter to Williams RJ, 17 May 1943. Roger Williams Papers, Box 88-087/1l, Folder: Williams, Roger J. – Personal Correspondence, etc., Dolph Briscoe Center for American History, The University of Texas at Austin.

Paterson DG. Letter to Williams RJ, 15 February 1957. Roger Williams Papers, Box 88-087/26a, Folder: Famous Names (Humanics), Dolph Briscoe Center for American History, The University of Texas at Austin.

Pauling L. Letter to Williams RJ, 6 March 1946. Roger Williams Papers, Box 88-087/26a, Folder: Correspondence Concerning Humanics Sept. 1945 – March 1946, Dolph Briscoe Center for American History, The University of Texas at Austin.

Raible R. Letter to Williams RJ, 26 October 1947. Roger Williams Papers, Box 88-087/26a, Folder: Comments from Important Men Regarding The Human Frontier, Dolph Briscoe Center for American History, The University of Texas at Austin.

Rhind FM. Letter to Painter TS, 20 May 1949. Roger Williams Papers, Box 88-087/1e, Folder: Rockefeller Foundation Cumulative, Dolph Briscoe Center for American History, The University of Texas at Austin.

Rockefeller JDI. Letter to Williams RJ, 24 April 1947. Roger Williams Papers, Box 88-087/26a, Folder: Famous Names (Humanics), Dolph Briscoe Center for American History, The University of Texas at Austin.

Rocker R. Letter to Williams RJ, 10 May 1943. Roger Williams Papers, Box 88-087/1l, Folder: Williams, Roger J. – Personal Correspondence, etc., Dolph Briscoe Center for American History, The University of Texas at Austin.

Rogers LL. Letter to Boman TJ, 9 June 1952. Roger Williams Papers, Box 88-087/23a, Folder: Alcoholism – Inquiries, Reports, Correspondence, Physicians, Scientists, Governmental Agencies, 1955 –, Dolph Briscoe Center for American History, The University of Texas at Austin.

Rogers LL, Williams RJ. Final Report For The Period 15 June 1949 – 14 May 1954. Austin, Texas: Biochemical Institute, The University of Texas, 1954. Roger Williams Papers, Box 88-087/20a, Folder: Naval Research, Dolph Briscoe Center for American History, The University of Texas at Austin.

Shivers A. Letter to Williams RJ, 13 June 1950. Roger Williams Papers, Box 88-087/41, Folder: LETTERS: RE "SHALL WE PIONEER TOO?", Dolph Briscoe Center for American History, The University of Texas at Austin.

Stafford J. Food Desires Guide Weight. World-Telegram, 11 August 1945. Roger Williams Papers, Box 88-087/1k, "What To Do About Vitamins", Dolph Briscoe Center for American History, The University of Texas at Austin.

Stassen HE. Letter to Williams RJ, 24 June 1946. Roger Williams Papers, Box 88-087/26a, Folder: Famous Names (Humanics), Dolph Briscoe Center for American History, The University of Texas at Austin.

Stettinius Jr. ER. Letter to Williams RJ, 25 October 1946. Roger Williams Papers, Box 88-087/26a, Folder: Famous Names (Humanics), Dolph Briscoe Center for American History, The University of Texas at Austin.

Texas Professor Heads American Chemical Group. Valley Star, 11 December 1955. Roger Williams Papers, Box 88-087/47b, Folder: Clippings, Dolph Briscoe Center for American History, The University of Texas at Austin.

Texas Professor Honored as Head Of Chemist Group. Times Review, 11 December 1955. Roger Williams Papers, Box 88-087/47b, Folder: Clippings, Dolph Briscoe Center for American History, The University of Texas at Austin.

The Editors. Roger J. Williams: Nutrition and alcoholism (Ernährung und Alkoholismus). Chem Zentralblatt 1953. Roger Williams Papers, Box 88-087/23a, Folder: Alcoholism – Inquiries, Reports, Correspondence, Physicians, Scientists, Governmental Agencies, 1955 –, Dolph Briscoe Center for American History, The University of Texas at Austin.

The Physician's Bookshelf. Ohio State Med J 1945: 690. Roger Williams Papers, Box 88-087/1k, "What To Do About Vitamins", Dolph Briscoe Center for American History, The University of Texas at Austin.

University of Oklahoma Press. Statement of Sales: What To Do About Vitamins 1950. Roger Williams Papers, Box 88-087/1l Folder: Dr Williams Personal Correspondence – 1950 –, Dolph Briscoe Center for American History, The University of Texas at Austin.

University of Oklahoma Press. Statement of Sales: What To Do About Vitamins 1953a. Roger Williams Papers, Box 88-087/47b, Folder: Royalties, Dolph Briscoe Center for American History, The University of Texas at Austin.

University of Oklahoma Press. Statement of Sales: Nutrition and Alcoholism 1953b. Roger Williams Papers, Box 88-087/47b, Folder: Royalties, Dolph Briscoe Center for American History, The University of Texas at Austin.

University of Texas Prof is Named New President of ACS. The Sun, 12 December 1955. Roger Williams Papers, Box 88-087/47b, Folder: Clippings, Dolph Briscoe Center for American History, The University of Texas at Austin.

US Chemists Name UT Prof President. American Statesman 11 December 1955: Roger Williams Papers, Box 88-087/47b, Folder: Clippings, Dolph Briscoe Center for American History, The University of Texas at Austin.

UT Biochemist To Head Society. Times Herald 11 December 1955: Roger Williams Papers, Box 88-087/47b, Folder: Clippings, Dolph Briscoe Center for American History, The University of Texas at Austin.

UT Professor Elected Head American Chemical Society. Houston Post 11 December 1955: Roger Williams Papers, Box 88-087/47b, Folder: Clippings, Dolph Briscoe Center for American History, The University of Texas at Austin.

Vitamins and Their Roles in Everyday Diet. Morning Star-Telegram 30 September 1945. Roger Williams Papers, Box 88-087/1k, "What To Do About Vitamins", Dolph Briscoe Center for American History, The University of Texas at Austin.

Weaver W. Letter to Williams RJ, 21 November 1945. Roger Williams Papers, Box 88-087/26a, Folder: Correspondence Concerning Humanics Sept. 1945 – March 1946, Dolph Briscoe Center for American History, The University of Texas at Austin.

Weaver W. Letter to Williams RJ, 28 January 1946. Roger Williams Papers, Box 88-087/26a, Folder: Correspondence Concerning Humanics Sept. 1945 – March 1946, Dolph Briscoe Center for American History, The University of Texas at Austin.

Weaver W. Letter to Williams RJ, 10 August 1954. Roger Williams Papers, Box 88-087/8, Folder: Warren Weaver – 1953–1955, Dolph Briscoe Center for American History, The University of Texas at Austin.

What To Do About Vitamins. Whats New Home Econ 1945a. Roger Williams Papers, Box 88-087/1k, "What To Do About Vitamins", Dolph Briscoe Center for American History, The University of Texas at Austin.

What To Do About Vitamins. Infantry J Wash DC 1945b. Roger Williams Papers, Box 88-087/1k, "What To Do About Vitamins", Dolph Briscoe Center for American History, The University of Texas at Austin.

What to do About Vitamins. J Am Med Assoc 1945. Roger Williams Papers, Box 88-087/1k, "What To Do About Vitamins", Dolph Briscoe Center for American History, The University of Texas at Austin.

What To Do About Vitamins. Med World 1946: 178. Roger Williams Papers, Box 88-087/1k, "What To Do About Vitamins", Dolph Briscoe Center for American History, The University of Texas at Austin.

Williams RJ. Letter to Harper and Brothers, 15 October 1945c. Roger Williams Papers, Box 88-087/26a, Folder: Correspondence Concerning Humanics Sept. 1945 – March 1946, Dolph Briscoe Center for American History, The University of Texas at Austin.

Williams RJ. Letter to de Kruif P, 2 November 1945f. Roger Williams Papers, Box 88-087/26a, Folder: Correspondence Concerning Humanics Sept. 1945 – March 1946, Dolph Briscoe Center for American History, The University of Texas at Austin.

Williams RJ. Notes on the Future of Research in Humanology 1945g. Roger Williams Papers, Box 88-087/41, Folder: Humanics… Unpublished, Dolph Briscoe Center for American History, The University of Texas at Austin.

Williams RJ. Letter to Wallace HA, 7 December 1945h. Roger Williams Papers, Box 88-087/26a, Folder: Correspondence Concerning Humanics Sept. 1945 – March 1946, Dolph Briscoe Center for American History, The University of Texas at Austin.

Williams RJ. Letter to Morse WL, 10 December 1945i. Roger Williams Papers, Box 88-087/26a, Folder: Correspondence Concerning Humanics Sept. 1945 – March 1946, Dolph Briscoe Center for American History, The University of Texas at Austin.

Williams RJ. Letter to Secretary to Dr. Warren Weaver, 21 January 1946b. Roger Williams Papers, Box 88-087/26a, Folder: Correspondence Concerning Humanics Sept. 1945 – March 1946, Dolph Briscoe Center for American History, The University of Texas at Austin.

Williams RJ. Letter to Seashore RH, 19 February 1946c. Roger Williams Papers, Box 88-087/26a, Folder: Correspondence Concerning Humanics Sept. 1945 – March 1946, Dolph Briscoe Center for American History, The University of Texas at Austin.

Williams RJ. Letter to de Kruif P, 2 January 1946d. Roger Williams Papers, Box 88-087/26a, Folder: Correspondence Concerning Humanics Sept. 1945 – March 1946, Dolph Briscoe Center for American History, The University of Texas at Austin.

Williams RJ. Letter to de Kruif P, 10 January 1946e. Roger Williams Papers, Box 88-087/26a, Folder: Correspondence Concerning Humanics Sept. 1945 – March 1946, Dolph Briscoe Center for American History, The University of Texas at Austin.

Williams RJ. Letter to de Kruif P, 7 March 1946f. Roger Williams Papers, Box 88-087/26a, Folder: Correspondence Concerning Humanics Sept. 1945 – March

1946, Dolph Briscoe Center for American History, The University of Texas at Austin.

Williams RJ. Letter to Keener JD, 29 October 1946g. Roger Williams Papers, Box 88-087/26a, Folder: Humanics: A Crucial Need, Dolph Briscoe Center for American History, The University of Texas at Austin.

Williams RJ. Letter to McKeithan DM, 20 May 1946h. Roger Williams Papers, Box 88-087/26a, Folder: Humanics Article Faculty Reactions, Dolph Briscoe Center for American History, The University of Texas at Austin.

Williams RJ. Letter to Valentine WL, 28 February 1946i. Roger Williams Papers, Box 88-087/26a, Folder: Correspondence Concerning Humanics Sept. 1945 – March 1946, Dolph Briscoe Center for American History, The University of Texas at Austin.

Williams RJ. Letter to Cole FE, 13 June 1946j. Roger Williams Papers, Box 88-087/1l, Folder: Williams, Roger J. – Personal Correspondence, etc., Dolph Briscoe Center for American History, The University of Texas at Austin.

Williams RJ. Letter to Campbell FL, 19 June 1946k. Roger Williams Papers, Box 88-087/26a, Folder: Humanics: A Crucial Need, Dolph Briscoe Center for American History, The University of Texas at Austin.

Williams RJ. Letter to Lal GB, 3 February 1947c. Roger Williams Papers, Box 88-087/47b, Folder: My Faith, Dolph Briscoe Center for American History, The University of Texas at Austin.

Williams RJ. Comments Concerning Roger J. Williams' Book and Washington Address Both Entitled, "The Human Frontier" 1947f. Roger Williams Papers, Box 88-087/1i, Folder: The Human Frontier (Comments), Dolph Briscoe Center for American History, The University of Texas at Austin.

Williams RJ. Comments From Medical Men 1947g. Roger Williams Papers, Box 88-087/26a, Folder: Comments from Important Men Regarding The Human Frontier, Dolph Briscoe Center for American History, The University of Texas at Austin.

Williams RJ. Comments Regarding "The Human Frontier" 1947h. Roger Williams Papers, Box 88-087/26a, Folder: Comments from Important Men Regarding The Human Frontier, Dolph Briscoe Center for American History, The University of Texas at Austin.

Williams RJ. Statements From Ministers Regarding "The Human Frontier" 1947i. Roger Williams Papers, Box 88-087/26a, Folder: Comments from Important Men Regarding The Human Frontier, Dolph Briscoe Center for American History, The University of Texas at Austin.

Williams RJ. Overview of Grants to The Department of Biochemistry at The University of Texas 1949a. Roger Williams Papers, Box 88-087/47b,

Folder: Williams Personell: Miscellaneous, Dolph Briscoe Center for American History, The University of Texas at Austin.

Williams RJ. A Plea For Temperance and Discrimination in the Generic Use of the Word "The". 1950c. Roger Williams Papers, Box 88-087/41, Folder: Plea for temperance and discrimination in the generic use of the word "the", Dolph Briscoe Center for American History, The University of Texas at Austin.

Williams RJ. Letter to Baughman JL, 7 December 1950d. Roger Williams Papers, Box 88-087/1i, Folder: Publications Various (Corres. Regarding, etc.), Dolph Briscoe Center for American History, The University of Texas at Austin.

Williams RJ. Introduction, General Discussion and Tentative Conclusions. Biochem. Inst. Stud. IV Individ. Metab. Patterns Hum. Dis. Explor. Study Util. Predominantly Pap. Chromatogr. Methods, Austin, Texas: The University of Texas, Austin, 1951b, 7–21.

Williams RJ. Letter to Reynolds PR, 1 October 1951f. Roger Williams Papers, Box 88-087/1i, Folder: Publishers Various, Dolph Briscoe Center for American History, The University of Texas at Austin.

Williams RJ. Letter to Wardlaw FH, 25 January 1952c. Roger Williams Papers, Box 88-087/1k, Folder: The University of Oklahoma Press (Lottinville) 1950–1951, Dolph Briscoe Center for American History, The University of Texas at Austin.

Williams RJ. Letter to Weaver W, 8 September 1954c. Roger Williams Papers, Box 88-087/8, Folder: Warren Weaver – 1953–1955, Dolph Briscoe Center for American History, The University of Texas at Austin.

Williams RJ. Letter to Weaver W, 30 May 1955b. Roger Williams Papers, Box 88-087/8, Folder: Warren Weaver – 1953–1955, Dolph Briscoe Center for American History, The University of Texas at Austin.

Williams RJ. Letter to Weaver W, 4 January 1955c. Roger Williams Papers, Box 88-087/8, Folder: Warren Weaver – 1953–1955, Dolph Briscoe Center for American History, The University of Texas at Austin.

Williams RJ. Letter to Russell B, 21 March 1957d. Roger Williams Papers, Box 88-087/26a, Folder: Famous Names (Humanics), Dolph Briscoe Center for American History, The University of Texas at Austin.

Williams RJ. Letter to Allen WT, 13 November 1958. Roger Williams Papers, Box 88-087/23a, Folder: Alcoholism Letters A, Dolph Briscoe Center for American History, The University of Texas at Austin.

Williams RJ. Titles, Undated. Roger Williams Papers, Box 88-087/41, Folder: Humanics ... Unpublished, Dolph Briscoe Center for American History, The University of Texas at Austin.

Williams RJ. Outline to Man, Frustrated or Free, Undated. Roger Williams Papers, Box 88-087/41, Folder: Man, Frustrated or Free, Dolph Briscoe Center for American History, The University of Texas at Austin.

Williams RJ, Beerstecher Jr. E, Berry LJ, LaBrosse EH. The Experimental Treatment of Genetotrophic Disease. Austin, Texas: 1950b. Roger Williams Papers, Box 88-087/20a, Folder: The Experimental Treatment of Genetotrophic Disease, Dolph Briscoe Center for American History, The University of Texas at Austin.

Williams RR. Letter to Williams RJ, 3 February 1946l. Roger Williams Papers, Box 88-087/26a, Folder: Correspondence Concerning Humanics Sept. 1945 – March 1946, Dolph Briscoe Center for American History, The University of Texas at Austin.

Williams RR. Letter to Williams RJ, 1946m. Roger Williams Papers, Box 88-087/26a, Folder: Correspondence Concerning Humanics Sept. 1945 – March 1946, Dolph Briscoe Center for American History, The University of Texas at Austin.

Wynne AG. Letter to Williams RJ, 12 June 1950. Roger Williams Papers, Box 88-087/41, Folder: LETTERS: RE "SHALL WE PIONEER TOO?", Dolph Briscoe Center for American History, The University of Texas at Austin.

Yerkes RM. Letter to Williams RJ, 16 June 1950. Roger Williams Papers, Box 88-087/41, Folder: LETTERS: RE "SHALL WE PIONEER TOO?", Dolph Briscoe Center for American History, The University of Texas at Austin.

19.2 Published Materials

Abbott EA. Francis Bacon: An Account of His Life and Works. Hamburg: Severus, 2013.

AbeBooks Inc., 2023: AbeBooks. https://www.abebooks.com/ (accessed 25 July 2023).

Abrahams E, Silver M. The History of Personalized Medicine. In: Gordon E, Koslow S, eds. Integr. Neurosci. Pers. Med., Oxford: Oxford University Press, 2011, 3–17.

Adams CE, Williams RJ. Laboratory Preparation of Acetaldehyde. J Am Chem Soc 1921; 43: 2420–2421. https://doi.org/10.1021/ja01444a015.

Ainsworth GC. Introduction to the History of Mycology. Cambridge: Cambridge University Press, 1976.

Alcoholics Anonymous World Services, Inc., 2017: This is A. A.: An Introduction to the A.A. Recovery Program. https://www.aa.org/aa-introduction-aa-recovery-program.

American Association for the Advancement of Science. Registration in the National Roster. Science 1942; 96: 292. https://doi.org/10.1126/science.96.2491.292.

American Medical Association, 2023: AMA adopts new policy clarifying role of BMI as a measure in medicine. https://www.ama-assn.org/press-center/press-releases/ama-adopts-new-policy-clarifying-role-bmi-measure-medicine (accessed 9 July 2023).

American Scientist, 2023: About Us. https://www.americanscientist.org/content/about-us (accessed 21 February 2023).

Anastasi A. Review of Free and Unequal: The Biological Basis of Individual Liberty. Hum Biol 1955; 27: 243–246.

Anderson JE. Genus Homo. Sci Mon 1947; 64: 443–445.

Anson BJ. An Atlas of Human Anatomy. Philadelphia: Saunders, 1950.

Applebaum A. Iron Curtain: The Crushing of Eastern Europe 1944–56. London: Penguin, 2012.

Arbeitsgemeinschaft der Wissenschaftlichen Medizinischen Fachgesellschaften eV, 2023: AWMF-Regelwerk Leitlinien. https://www.awmf.org/regelwerk/ (accessed 19 February 2023).

Arey LB. Barry J. Anson. A Biographical Sketch. Q Bull Northwest Univ Evanst Ill Med Sch 1962; 36: 185–188.

Asquith GH. Anton T. Boisen and the Study of "Living Human Documents". J Presbyt Hist 1982; 60: 244–265.

Badash L. Science and McCarthyism. Minerva 2000; 38: 53–80.

Badrick T. Biological variation: Understanding Why It Is So Important? Pract Lab Med 2021; 23: e00199. https://doi.org/10.1016/j.plabm.2020.e00199.

Ballou RB. Critical Crossroads. Teach Coll Rec 1955; 56: 309–315.

Baumann LC, Karel A. Health Education. In: Gellman MD, Turner JR, eds. Encycl. Behav. Med., New York, NY: Springer, 2013, 917–918. https://doi.org/10.1007/978-1-4419-1005-9_320.

Beadle GW. Biochemical Genetics. Chem Rev 1945; 37: 15–96. https://doi.org/10.1021/cr60116a002.

Beadle GW. Genes and Chemical Reactions in Neospora. Nobel Lect 1958: 588–599.

Beadle GW, Tatum EL. Genetic Control of Biochemical Reactions in Neurospora. Proc Natl Acad Sci U S A 1941; 27: 499–506.

Bean WB. Biochemical Individuality: The Basis for the Genetotrophic Concept. AMA Arch Intern Med 1957; 100: 342–343. https://doi.org/10.1001/archinte.1957.00260080168049.

Beerstecher, Ernest, Jr.: Biochemistry. Am Men Women Sci 1989–90 1989; 1: 406.

Beerstecher, Ernest, Jr.: Biochemistry. Am Men Women Sci 1998–99 1998; 1: 496.

Beerstecher Jr. E, Reed J, Brown WD, Berry LJ. The Effects of Single Vitamin Deficiencies on the Consumption of Alcohol by White Rats. Biochem. Inst. Stud. IV Individ. Metab. Patterns Hum. Dis. Explor. Study Util. Predominantly Pap. Chromatogr. Methods, Austin, Texas: The University of Texas, Austin, 1951, 115–138.

Beerstecher Jr. E, Sutton HE, Berry HK, Brown WD, Reed J, Rich GB, Berry LJ, Williams RJ. Biochemical Individuality. V. Explorations with Respect to the Metabolic Patterns of Compulsive Drinkers. Arch Biochem 1950; 29: 27–40.

Beloff RH. Biochemical Individuality. Yale J Biol Med 1957; 29: 557–558.

Bennett AE. Biochemical Individuality. The Basis for the Genetotrophic Concept. Am J Psychiatry 1958: 1054–1055. https://doi.org/10.1176/ajp.114.11.1054-b.

Berry HK, Cain L. Biochemical Individuality IV: A Paper Chromatographic Technique for Determining Excretion of Amino Acids in the Presence of Interfering Substances. Arch Biochem 1949; 24: 179–189.

Bhatt A. Evolution of Clinical Research: A History Before and Beyond James Lind. Perspect Clin Res 2010; 1: 6–10.

Bibbins-Domingo K, Curfman G, 2023: About JAMA. https://jamanetwork.com/journals/jama/pages/for-authors#fa-about (accessed 21 February 2023).

Biesele JJ. In Memoriam: Roger J. Williams – 1893–1988. J Orthomol Med 1988; 3: 53–54.

Biochemical Individuality. The Basis for the Genetotrophic Concept. AIBS Bull 1957; 7: 43. https://doi.org/10.2307/1292230.

Biochemist Roger Williams dead at 94. Chem Eng News Arch 1988; 66: 66. https://doi.org/10.1021/cen-v066n010.p066.

Bland JS. The Evolution of Personalized Nutrition – From Addis, Pauling, and RJ Williams to the Future. Integr Med Clin J 2019a; 18: 10–13.

Bland JS. Systems Biology Meets Functional Medicine. Integr Med Clin J 2019b; 18: 14–18.

Blum D. Medical Group Says B.M.I. Alone Is Not Enough to Assess Health and Weight. The New York Times 15 June 2023: .

Boisen AT. The Human Frontier. By Roger J. Williams. J Relig 1947; 27: 297–298. https://doi.org/10.1086/483633.

Boyd WC. Biochemical Individuality. Am J Hum Genet 1957; 9: 141–142.

Bracken MB. Why Animal Studies Are Often Poor Predictors of Human Reactions to Exposure. J R Soc Med 2009; 102: 120–122. https://doi.org/10.1258/jrsm.2008.08k033.

Briscoe Center for American History, 2023: AEON. https://briscoecenter.aeon.atlas-sys.com/aeonauth/aeon.dll?Action=10&Form=10 (accessed 25 July 2023).

Bull JP. A Study of the History and Principles of Clinical Therapeutic Trials. PhD Thesis. University of Cambridge, 1951.

Butt JE, 2023: Alexander Pope. https://www.britannica.com/biography/Alexander-Pope-English-author (accessed 21 February 2023).

Cammack RC, Atwood TA, Campbell PC, Parish HP, Smith AS, Vella FV, Stirling JS. Polar Coordinate. Oxf. Dict. Biochem. Mol. Biol., Oxford: Oxford University Press, 2006.

Candlish J. Biochemical individuality: By R J Williams, pp 214. John Wiley & Sons, 1956. Biochem Educ 1989; 17: 54. https://doi.org/10.1016/0307-4412(89)90081-2.

Carmichael L. Karl Spencer Lashley, Experimental Psychologist. Science 1959; 129: 1410–1412. https://doi.org/10.1126/science.129.3360.1410.

Catalano C. Shaping the American Woman: Feminism and Advertising in the 1950s. Constr Past 2002; 3.

Cattell J. American Men of Science. 7th ed. Lancaster, Pennsylvania: The Science Press, 1944.

Cesario A, D'Oria M, Auffray C, Scambia G. Personalized Medicine meets Artificial Intelligence: Beyond "Hype", Towards the Metaverse. 1st ed. 2023 Edition. Cham: Springer, 2023.

Chemical & Engineering News, 2023: About Us. https://cen.acs.org/static/about/aboutus.html (accessed 21 February 2023).

Chemical Abstracts Service, 2023: Our CAS Story – Connections That Matter. http://www.cas.org/about (accessed 21 February 2023).

Chillakuri B, Mahanandia R. Generation Z Entering the Workforce: The Need for Sustainable Strategies in Maximizing Their Talent. Hum Resour Manag Int Dig 2018; 26: 34–38. https://doi.org/10.1108/HRMID-01-2018-0006.

City of Austin, 2009: Zoning Change Review Sheet: The Roger Williams House. https://www.austintexas.gov/edims/document.cfm?id=132351 (accessed 22 February 2023).

Clapham C, Nicholson J. Polar Coordinates. Concise Oxf. Dict. Math. Oxford: Oxford University Press, 2009.

Clarke L, 2022: "It's a modern-day Facebook" – how BeReal became Gen Z's favourite app. https://www.theguardian.com/media/2022/aug/21/its-a-modern-day-facebook-how-bereal-became-gen-zs-favourite-app (accessed 8 February 2023).

Clayton Biotechnologies, 2023: Clayton Biotechnologies. https://www.claytonbiotech.com/ (accessed 21 February 2023).

Cochrane Deutschland, 2023: Leitlinien. https://www.cochrane.de/leitlinien (accessed 19 February 2023).

Cohen R, Newton-John T, Slater A. The Case for Body Positivity on Social Media: Perspectives on Current Advances and Future Directions. J Health Psychol 2021; 26: 2365–2373. https://doi.org/10.1177/1359105320912450.

Collier R. Legumes, Lemons and Streptomycin: A Short History of the Clinical Trial. CMAJ Can Med Assoc J 2009; 180: 23–24. https://doi.org/10.1503/cmaj.081879.

Comfort NC. The Science of Human Perfection: How Genes Became the Heart of American Medicine. New Haven: Yale University Press, 2012.

Considine GD, editor. Van Nostrand's Scientific Encyclopedia. New York: John Wiley & Sons, 2005. https://doi.org/10.1002/9780471743989.

Cooper JM, 2023: Theodore Roosevelt. https://www.britannica.com/biography/Theodore-Roosevelt (accessed 21 February 2023).

Cotlier E. Dangers of Vitamin Therapy for Senile Cataracts. N Engl J Med 1977; 296: 398–399. https://doi.org/10.1056/NEJM197702172960718.

Council on Pharmacy and Chemistry. J Am Med Assoc 1949; 140: 534–539. https://doi.org/10.1001/jama.1949.02900410030008.

Courtney AE, Lockeretz SW. A Woman's Place: An Analysis of the Roles Portrayed by Women in Magazine Advertisements. J Mark Res 1971; 8: 92–95. https://doi.org/10.2307/3149733.

Cowles C, 2022: Can "Body Neutrality" Change the Way You Work Out? https://www.nytimes.com/2022/02/02/well/move/body-neutrality-exercise.html (accessed 8 February 2023).

Craft AW, Editors T. The First Randomised Controlled Trial. Arch Dis Child 1998; 79: 410. https://doi.org/10.1136/adc.79.5.410.

Danforth CH. Free and Unequal: The Biological Basis of Individual Liberty. Am J Hum Genet 1953; 5: 402–404.

Daniels LA, 1992: Alfred McClung Lee Dies at 85; Professor Was Noted Sociologist. https://www.nytimes.com/1992/05/21/nyregion/alfred-mcclung-lee-dies-at-85-professor-was-noted-sociologist.html (accessed 22 October 2020).

Davidson DL. What to Do About Vitamins (Williams, Roger J.). J Chem Educ 1945; 22: 416. https://doi.org/10.1021/ed022p416.3.

Davis DR. Nutritional Prevention of Cataracts. N Engl J Med 1978; 298: 55. https://doi.org/10.1056/NEJM197801052980119.

Davis DR. In Memoriam: Roger John Williams 1893–1988. J Appl Nutr 1988; 40: 121–125.

Davis DR, 2003a: Roger J. Williams, Pioneer in Nutrition and Biochemical Individuality. http://bioinst.cm.utexas.edu/williams/index.html (accessed 21 February 2023).

Davis DR, 2003b: Articles 1919-1944, #1-108. http://bioinst.cm.utexas.edu/williams/1919.htm (accessed 21 February 2023).

Davis DR, 2003c: Articles 1945-1964, #109-209. http://bioinst.cm.utexas.edu/williams/1945.htm (accessed 21 February 2023).

Davis DR, 2003d: Articles 1965 & Later, #210–278. http://bioinst.cm.utexas.edu/williams/1965.htm (accessed 21 February 2023).

Davis DR, 2003e: Books by Roger J. Williams. http://bioinst.cm.utexas.edu/williams/books.htm (accessed 21 February 2023).

Davis DR, 2010: Williams, Roger John. http://www.tshaonline.org/handbook/online/articles/fwiav (accessed 21 February 2023).

Davis DR, Hackert ML, Reed LJ, 2008: Roger J. Williams 1893–1988 A Biographical Memoir. http://www.nasonline.org/publications/biographical-memoirs/memoir-pdfs/williams-roger.pdf (accessed 7 May 2023).

Dempsey EW. New Book. Endocrinology 1946; 39: 432–433. https://doi.org/10.1210/endo-39-6-432.

Department of Education and Training Victoria, 2023: Individual Education Plans (IEPs). https://www2.education.vic.gov.au/pal/individual-education-plans-ieps/undefined (accessed 3 March 2023).

Deutsche Nationalbibliothek, 2023: Zeitschriften Datenbank. https://zdb-katalog.de/index.xhtml (accessed 25 July 2023).

Dietotherapy: Clinical Application of Modern Nutrition. Acad Med 1945; 20: 271.

Draper G. The Relationship of Human Constitution to Disease. Science 1925; 61: 525–528. https://doi.org/10.1126/science.61.1586.525.

Draper G, Dupertuis CW, Caughey JL. Human Constitution in Clinical Medicine. New York: Paul B. Hoeber, Inc., 1944.

Dunitz JD, 1997: Linus Carl Pauling. http://www.nasonline.org/publications/biographical-memoirs/memoir-pdfs/pauling-linus.pdf (accessed 23 February 2023).

Eli Lilly and Company, 2023: Key Facts. https://www.lilly.com/who-we-are/about-lilly/key-facts (accessed 21 February 2023).

Elsevier B.V., 2023: SCOPUS. https://www.scopus.com/search/form.uri?display=basic#basic (accessed 25 July 2023).

Emmert-Streib F. Personalized Medicine: Has It Started Yet? A Reconstruction of the Early History. Front Genet 2013; 3: 1–4. https://doi.org/10.3389/fgene.2012.00313.

Eppright MA, Williams RJ. Thiamine Determination Comparative Study of Yeast-Growth, Yeast-Fermentation, and Thiochrome Methods. Ind Eng Chem Anal Ed 1944; 16: 576–579. https://doi.org/10.1021/i560133a012.

European Commission, 2023: Personalised medicine. https://health.ec.europa.eu/medicinal-products/personalised-medicine_en (accessed 5 October 2023).

Fitzgerald K, Rountree R. Biological Age and Functional Medicine: A Clinical Conversation with Kara Fitzgerald, ND, IFMCP, and Robert Rountree, MD. Integr Complement Ther 2022. 140 Huguenot Street, 3rd Floor New Rochelle, NY 10801 USA. https://doi.org/10.1089/ict.2022.29030.kfi.

Fleming R. Medical Treatment of the Inebriate. Q J Stud Alcohol 1945; 6: 391.

Fowler G, 1988: Roger J. Williams Is Dead at 94; Biochemist and Nutrition Expert. https://www.nytimes.com/1988/02/23/obituaries/roger-j-williams-is-dead-at-94-biochemist-and-nutrition-expert.html (accessed 19 June 2019).

Francis A. Countway Library of Medicine, Center for the History of Medicine, 2023: Heath, Clark Wright, 1900-. Papers, 1928-1955 (inclusive): Finding Aid. https://hollisarchives.lib.harvard.edu/repositories/14/resources/6597 (accessed 21 February 2023).

Francis T, Hoefel F, 2018: True Gen': Generation Z and its implications for companies. https://www.mckinsey.com/industries/consumer-packaged-goods/our-insights/true-gen-generation-z-and-its-implications-for-companies (accessed 23 February 2023).

Free and Unequal: The Biological Basis of Individual Liberty. Am Inst Biol Sci Bull 1953; 3: 13. https://doi.org/10.1093/aibsbulletin/3.3.13.

Gadebusch Bondio M. "Das Individuum – eine Abweichung" … und das Unbehagen der Wissenschaft. In: Brinkschulte E, Gadebusch Bondio M, eds. Norm Als Zwang Pflicht Traum Normierende Individ. Bestrebungen Med., vol. 19, Frankfurt am Main, Berlin, Bern, Bruxelles, New York, Oxford, Wien: Peter Lang, 2015, 19–50.

Gadebusch Bondio M. Beyond the Causes of Disease: Prediction and the Need for a New Philosophy of Medicine. In: Spöring F, Gadebusch Bondio M,

Gordon J-S, eds. Med. Ethics Predict. Progn. 1st ed., New York: Routledge, 2017, 11–29.

Gadebusch Bondio M, Spöring F. Personalized Medicine: Historical Roots of a Medical Model. In: Boniolo G, Nathan MJ, eds. Philos. Mol. Med. Found. Issues Res. Pract., New York: Routledge, 2017, 35–56.

Galen. A Translation Of Galen's Hygiene: De Sanitate Tuenda. Literary Licensing, LLC, 2012.

Gamber W, 2019: Women and Domesticity in the 1950s. https://oxfordre.com/americanhistory/display/10.1093/acrefore/9780199329175.001.0001/acrefore-9780199329175-e-423?rskey=ZQxtMW&result=9 (accessed 23 February 2023).

Garn SM, Giles E. Earnest Albert Hooton (1887–1954). Biogr. Mem., Washington, D.C.: National Academies Press, 1995, 167–179, National Academy of Sciences Online.

Garrod AE. The Incidence of Alkaptonuria: A Study in Chemical Individuality. Mol Med 1902; 2: 274–282.

Garrod AE. Inborn Errors of Metabolism. 2nd ed. London: Oxford University Press, 1923.

Garside EB, 1947: Pointers for Improving Man's Knowledge of Himself. https://www.nytimes.com/1947/01/26/archives/pointers-for-improving-mans-knowledge-of-himself-the-human-frontier.html (accessed 23 February 2023).

Gartler SM, Jarvik GC, 2019: Arno G. Motulsky 1923 – 2018. http://www.nasonline.org/publications/biographical-memoirs/memoir-pdfs/motulsky-arno.pdf (accessed 27 January 2023).

Gasta M. The Nutrition Assessment of Metabolic and Nutritional Balance. In: Noland D, Drisko JA, Wagner L, eds. Integr. Funct. Med. Nutr. Ther. Princ. Pract., Cham: Springer International Publishing, 2020, 99–122. https://doi.org/10.1007/978-3-030-30730-1_8.

Gayon J. From Mendel to Epigenetics: History of genetics. C R Biol 2016; 339: 225–230. https://doi.org/10.1016/j.crvi.2016.05.009.

Giera M, Yanes O, Siuzdak G. Metabolite Discovery: Biochemistry's Scientific Driver. Cell Metab 2022; 34: 21–34. https://doi.org/10.1016/j.cmet.2021.11.005.

Glass B. What to Do About Vitamins. Roger J. Williams. Q Rev Biol 1946; 21: 202. https://doi.org/10.1086/395270.

Glass B. Review of the Human Frontier. A New Pathway for Science Toward a Better Understanding of Ourselves. Q Rev Biol 1947; 22: 175–176.

Goforth AN, Pham AV, Carlson JS. Diathesis-stress Model. In: Goldstein S, Naglieri JA, eds. Encycl. Child Behav. Dev., Boston, MA: Springer US, 2011, 502–503. https://doi.org/10.1007/978-0-387-79061-9_845.

Gonzalez MJ, Massari JRM. Metabolic Correction : A Functional Explanation of Orthomolecular Medicine. J Orthomol Med 2012; 27.

Good CV. Research Methods Biliography. Phi Delta Kappan 1947; 29: 146–152.

Goodpasture EW, 1961: Leo Loeb. http://www.nasonline.org/publications/biogr aphical-memoirs/memoir-pdfs/loeb-leo.pdf (accessed 13 October 2020).

Google, 2023: Google Scholar. https://scholar.google.de/ (accessed 26 July 2023).

Greenstein JP. Biochemical individuality: The basis for the genetotrophic concept: By Roger J. Williams. John Wiley, New York, New York, 1956. xiv + 214 pp. Price $5.75. Arch Biochem Biophys 1957; 70: 633–634. https://doi.org/10.1016/0003-9861(57)90160-1.

Grenell RG. Biochemical Individuality. The Basis for the Genetotrophic Concept. Roger J. Williams. Q Rev Biol 1957; 32: 306. https://doi.org/10.1086/401939.

Guyatt G, Cairns J, Churchill D, Cook D, Haynes B, Hirsh J, Irvine J, Levine M, Levine M, Nishikawa J, Sackett D, Brill-Edwards P, Gerstein H, Gibson J, Jaeschke R, Kerigan A, Neville A, Panju A, Detsky A, Enkin M, et al. Evidence-Based Medicine: A New Approach to Teaching the Practice of Medicine. JAMA 1992; 268: 2420–2425. https://doi.org/10.1001/jama.1992.03490170092032.

György P, Rose CS, Eakin RE, Snell EE, Williams RJ. Egg-White Injury as the Result of Nonabsorption or Inactivation of Biotin. Science 1941; 93: 477–478. https://doi.org/10.1126/science.93.2420.477.

Hac LR, Snell EE, Williams RJ. The Microbiological Determination of Amino Acids II: Assay and Utilization of Glutamic Acid and Glutamine by Lactobacillus Arabinosus. J Biol Chem 1945; 159: 273–289.

Hackam DG, Redelmeier DA. Translation of Research Evidence From Animals to Humans. JAMA 2006; 296: 1727–1732. https://doi.org/10.1001/jama.296.14.1731.

Haight GS, 2023: George Eliot. https://www.britannica.com/biography/George-Eliot (accessed 21 February 2023).

Hallett LT. Obituary: Walter J. Murphy 1899–1959. Anal Chem 1960; 32: 1. https://doi.org/10.1021/ac60157a609.

Hanke ME. Fred Conrad Koch 1876–1948. Science 1948; 107: 671–672. https://doi.org/10.1126/science.107.2791.671.

Harrell MC, Bradley MA. Data Collection Methods: Semi-Structured Interviews and Focus Groups. RAND Corporation, 2009.

Harris RK. Biochemical Individuality: The Basis for the Genetotrophic Concept. Roger J. Williams John Wiley and Sons, Inc., New York, 1956. Pp. 214. $5.75. Philos Sci 1958; 25: 140–141. https://doi.org/10.1086/287590.

Hart CW. Franklin Delano Roosevelt: A Famous Patient. J Relig Health 2014; 53: 1102–1111. https://doi.org/10.1007/s10943-014-9886-3.

Hastings M. The Korean War: An Epic Conflict 1950–1953. Main Market edition. London: Pan, 2020.

Hawley RE, 2022: BeReal and the Fantasy of an Authentic Online Life. https://www.newyorker.com/culture/rabbit-holes/bereal-and-the-fantasy-of-an-authentic-online-life (accessed 8 February 2023).

Heath CW, Brouha L, Gregory LW, Seltzer CC, Wells FL, Woods WL. What People are: A Study of Normal Young Men. Cambridge, Massachusetts: Harvard University Press, 1945.

Heath CW, Brouha L, Gregory LW, Seltzer CC, Wells FL, Woods WL. What People Are: A Study of Normal Young Men. Second Printing. Cambridge, Massachusetts: Harvard University Press, 1946.

Hevesi D, 2009: Lorene Rogers, President of University of Texas in '70s, Is Dead at 94. https://www.nytimes.com/2009/01/26/education/26rogers.html (accessed 7 June 2021).

Hippokrates. Ausgewählte Schriften. De Gruyter (A), 2014. https://doi.org/10.1515/9783110361322.

Honigmann IG. John Joseph Honigmann. Anthropol Humanism Q 1982; 7: 10–13. https://doi.org/10.1525/ahu.1982.7.2-3.10.

Honigmann JJ. The Human Frontier. By Roger J. Williams. Harcourt Brace and Co., New York, viii + 314 pp., references, index, 1946. ($3.00). Am J Phys Anthropol 1947; 5: 378–379. https://doi.org/10.1002/ajpa.1330050311.

Hooton EA. Backside. Hum. Front. New Pathw. Sci. Better Underst. Ourselves, New York: Harcourt, Brace and Company, 1946.

Horn TD. Free and Unequal. Phi Delta Kappan 1954; 36: 93–98.

Horwitz AV. The Logic of Social Control. New York, NY: Springer Science & Business Media, 1990. https://doi.org/10.1007/978-1-4899-2230-4.

Howick J, Glasziou P. Philosophy of Evidence-based Medicine. 1. Edition. Chichester, West Sussex, UK: Wiley-Blackwell, 2011.

Hurley KE, Williams RJ. Urinary Amino Acids, Creatinine and Phosphate in Muscular Dystrophy. Arch Biochem Biophys 1955; 54: 384–391. https://doi.org/10.1016/0003-9861(55)90051-5.

Jacob F. The Logic of Life. 1st Edition. Princeton, N.J: Princeton University Press, 1993.

Jain KK. History of Medical Concepts Relevant to Personalized Medicine. Textb. Pers. Med., New York: Springer-Verlag, 2009, 3–5.

Jain KK. Textbook of Personalized Medicine. Cham: Springer International Publishing, 2021. https://doi.org/10.1007/978-3-030-62080-6.

Jenkins JP, 2022: White Supremacy. https://www.britannica.com/topic/white-supremacy (accessed 27 February 2023).

Johnson R, Hand L. IFMNT NIBLETS Nutrition Assessment Differential. In: Noland D, Drisko JA, Wagner L, eds. Integr. Funct. Med. Nutr. Ther. Princ. Pract., Cham: Springer International Publishing, 2020, 123–133. https://doi.org/10.1007/978-3-030-30730-1_9.

Jones DS. How Personalized Medicine Became Genetic, and Racial: Werner Kalow and the Formations of Pharmacogenetics. J Hist Med Allied Sci 2013a; 68: 1–48. https://doi.org/10.1093/jhmas/jrr046.

Jones WP. The March on Washington: Jobs, Freedom, and the Forgotten History of Civil Rights. Reprint Edition. New York City: W. W. Norton & Company, 2013b.

Kaku M, 2023: Albert Einstein. https://www.britannica.com/biography/Albert-Einstein (accessed 21 February 2023).

Kalow W. Pharmacogenetics: Heredity and the Response to Drugs. W.B. Saunders Company, 1962.

Kandutsch AA. Biochemical Individuality. J Hered 1957; 48: 168. https://doi.org/10.1093/oxfordjournals.jhered.a106713.

Keegan J. The Second World War. Reprint. New York: Penguin Books, 2005.

Kelly CC, Rhodes R. Manhattan Project: The Birth of the Atomic Bomb in the Words of Its Creators, Eyewitnesses, and Historians. New York City: Black Dog & Leventhal, 2009.

Kendler KS. A Prehistory of the Diathesis-Stress Model: Predisposing and Exciting Causes of Insanity in the 19th Century. Am J Psychiatry 2020; 177: 576–588. https://doi.org/10.1176/appi.ajp.2020.19111213.

Kety SS. Biochemical Individuality. The basis for the genetotrophic concept. Roger J. Williams. Wiley, New York; Chapman & Hall, London, 1956. 214 pp. Illus. $5.75. Science 1957; 125: 940. https://doi.org/10.1126/science.125.3254.940.a.

Keys A. Biochemical Individuality – The Basis for the Genetotrophic Concept. Am J Public Health Nations Health 1957; 47: 767. https://doi.org/10.2105/AJPH.47.6.767-a.

King JC. Review of Biochemical Individuality: The Basis for the Genetotrophic Concept. Eugen Q 1957; 4: 164–165.

Kirby Berry H, Cain L. Quantitative Study of Urinary and Salivary Amino Acids Using Paper Chromatography. Biochem. Inst. Stud. IV Individ. Metab. Patterns Hum. Dis. Explor. Study Util. Predominantly Pap. Chromatogr. Methods, Austin, Texas: The University of Texas, Austin, 1951, 71–76.

Kirby Berry H, Sutton HE, Cain L, Berry JS. Development of Paper Chromatography for Use in the Study of Metabolic Patterns. Biochem. Inst. Stud. IV Individ. Metab. Patterns Hum. Dis. Explor. Study Util. Predominantly Pap. Chromatogr. Methods, Austin, Texas: The University of Texas, Austin, 1951, 22–55.

Klarman MJ. From Jim Crow to Civil Rights: The Supreme Court and the Struggle for Racial Equality. Illustrated Edition. Oxford: Oxford University Press, U.S.A., 2006.

Kluger R. Simple Justice: The History of Brown v. Board of Education and Black America's Struggle for Equality. Revised Edition. New York, NY: Vintage, 2004.

Kohler RE. The History of Biochemistry: A Survey. J Hist Biol 1975; 8: 275–318.

Kojalo P, 2007: Houghton Mifflin Company Completes Acquisition of Harcourt Education, Harcourt Trade and Greenwood-Heinemann Divisions from Reed Elsevier, Creating Preeminent K–12 Educational Publisher. https://web.archive.org/web/20120210152808/http://www.hmco.com/company/investors/invest/ir_release_121307.html (accessed 18 August 2020).

Kraut R, 2022: Socrates. https://www.britannica.com/biography/Socrates (accessed 21 February 2023).

Krebs A, 1971: Dr. Paul de Kruif, Popularizer Of Medical Exploits, Is Dead. https://www.nytimes.com/1971/03/02/archives/dr-paul-de-kruif-popularizer-of-medical-exploits-is-dead.html (accessed 20 January 2023).

Kuhn TS, Hacking I. The Structure of Scientific Revolutions. 50th Anniversary Edition. Chicago: University of Chicago Press, 2012.

Lampe D. Roger J. Williams: Inside Individuality. Sciquest 1982: 22–24.

de Las Hazas M-CL, Dávalos A. Individualization, Precision Nutrition Developments for the 21st Century. In: Haslberger AG, ed. Adv. Precis. Nutr. Pers. Healthy Aging, Cham: Springer International Publishing, 2022, 25–50. https://doi.org/10.1007/978-3-031-10153-3_2.

Laurence WI. Expert Speaks: Put in Mold, Human Atom Goes Crazy. The Globe and Mail: Canada's National Newspaper 18 September 1948: 10.

Laxova R. Lionel Sharples Penrose, 1898–1972: A Personal Memoir in Celebration of the Centenary of His Birth. Genetics 1998; 150: 1333–1340.

Lee AM. Williams, Roger J. "The Human Frontier" (Book Review). Ann Am Acad Pol Soc Sci 1947; 249: 211.

Lewis M, Clark W, Ambrose SE. The Journals of Lewis and Clark. Revised Edition. Boston: Mariner Books, 1997.

Livermore AH. Biochemical Individuality (Williams, Roger J.). J Chem Educ 1958; 35: 373. https://doi.org/10.1021/ed035p373.5.

Loeb L. The Biological Basis of Individuality. 1st ed. Springfield, Illinois: Charles C Thomas, 1945.

Lowry L. The Giver. Reissue edition. New York: Ember, 1993.

Loyal Order of Moose, 2024: About Us. https://www.mooseintl.org/about-us/ (accessed 08 November 2024).

Machin D, Campbell M, Fayers P, Pinol A. Sample Size Tables for Clinical Studies. 2nd ed. Oxford, London, Berlin: Blackwell Science Ltd., 1987.

Marchand LA, 2023: Lord Byron. https://www.britannica.com/biography/Lord-Byron-poet (accessed 21 February 2023).

Maren TH, 1951: The Problem of "One More". https://www.nytimes.com/1951/09/09/archives/the-problem-of-one-more.html (accessed 4 April 2021).

Marshall A. Laying the foundations for personalized medicines. Nat Biotechnol 1997; 15: 4.

Martin BL. From Negro to Black to African American: The Power of Names and Naming. Polit Sci Q 1991; 106: 83–107. https://doi.org/10.2307/2152175.

Martin D, 2005: H. Bentley Glass, Provocative Science Theorist, Dies at 98 (Published 2005). https://www.nytimes.com/2005/01/20/science/h-bentley-glass-provocative-science-theorist-dies-at-98.html (accessed 23 February 2023).

Martin GJ. Biological relativity: The concept as it relates to chemotherapy and pharmacology. J Chem Educ 1956; 33: 204. https://doi.org/10.1021/ed033p204.

McCabe BF. Barry J. Anson. Ann Otol Rhinol Laryngol 1975; 84: 131. https://doi.org/10.1177/000348947508400121.

McCay CM. Review of Biochemical Individuality: the Basis for the Genetotrophic Concept. Am J Psychol 1957; 70: 672–673. https://doi.org/10.2307/1419483.

McIntosh C. The Iron Curtain. Camb. Adv. Learn. Dict. 4. Edition, Cambridge: Cambridge University Press, 2013.

Mendel G, 1866: Versuche über Pflanzen-Hybriden. https://library.si.edu/digital-library/book/versucheberpflan00mend (accessed 21 February 2023).

Merriam-Webster, 2023: Nosology. https://www.merriam-webster.com/dictionary/nosology (accessed 25 January 2023).

Mervin DM. McCarthyism. Concise Oxf. Dict. Polit. Int. Relat., Oxford: Oxford University Press, 2018.

Michaels J. McCarthyism: The Realities, Delusions and Politics Behind the 1950s Red Scare. New York: Taylor & Francis Ltd, 2017.

Michl S. The Epistemics of "Personalized Medicine". Rebranding Pharmacogenetics. In: Fischer T, Langanke M, Marschall P, Michl S, eds. Individ. Med. Ethical Econ. Hist. Perspect., Springer International Publishing, 2015, 61–78.

Mitchell HK, Houlahan MB. Adenine-Requiring Mutants of Neurospora Crassa. Fed Proc 1946; 5: 370–375.

Mitchell HK, Williams RJ. The Importance of Amino-Acids as Yeast Nutrients. Biochem J 1940; 34: 1532–1536.

Mitchell HK, Williams RJ. Folic Acid. III. Chemical and Physiological Properties. J Am Chem Soc 1944; 66: 271–274. https://doi.org/10.1021/ja01230a034.

Mitoma H, Manto M, Shaikh AG. Mechanisms of Ethanol-Induced Cerebellar Ataxia: Underpinnings of Neuronal Death in the Cerebellum. Int J Environ Res Public Health 2021; 18: 8678. https://doi.org/10.3390/ijerph18168678.

Montagu A. 1953: We Aren't the Same. https://www.nytimes.com/1953/05/17/archives/we-arent-the-same-free-and-unequal-the-biological-basis-of.html (accessed 4 April 2021).

Moore HE. The Human Frontier. By Roger J. Williams. New York: Harcourt, Brace and Company, 1946. 313 pp. $3.00. Soc Forces 1947; 26: 237–238. https://doi.org/10.2307/2571796.

Mosher WA, Saunders DH, Kingery LB, Williams RJ. Nutritional Requirements of the Pathogenic Mold Trichophyton Interdigitale. Plant Physiol 1936; 11: 795–806.

Motulsky AG. From Pharmacogenetics and Ecogenetics to Pharmacogenomics. Med Secoli 2002; 14: 683–705.

Motulsky AG. History of Human Genetics. In: Speicher MR, Motulsky AG, Antonarakis SE, eds. Vogel Motulskys Hum. Genet., Berlin, Heidelberg: Springer, 2010, 13–29. https://doi.org/10.1007/978-3-540-37654-5_2.

Murphy WJ. Science and Human Behavior. Chem Eng News Arch 1950; 28: 4529. https://doi.org/10.1021/cen-v028n052.p4529.

N. SM. Review of A New Pathway for Science. Phylon 1947; 8: 191–192, JSTOR. https://doi.org/10.2307/271730.

National Center for Biotechnology Information, 2023: PubMed. https://pubmed.ncbi.nlm.nih.gov/ (accessed 25 July 2023).

National Council on Family Relations. Review of The Human Frontier. Marriage Fam Living 1948; 10: 78, JSTOR. https://doi.org/10.2307/348982.

National Health Service, 2023: Coronavirus – Post-COVID syndrome (long COVID). https://www.england.nhs.uk/coronavirus/post-covid-syndrome-long-covid/ (accessed 26 July 2023).

National Institute for Health and Care Excellence, 2023a: Glossary. https://www.nice.org.uk/glossary?letter=r (accessed 21 February 2023).

National Institute for Health and Care Excellence, 2023b: About. https://www.nice.org.uk/about (accessed 21 February 2023).

Neustadt J, Pieczenik S. Biochemical Individuality. Integr Med 2007; 6: 30–32.

New Netherland Institute, 2023: David van Nostrand. https://www.newnetherlandinstitute.org/history-and-heritage/dutch_americans/david-van-nostrand/ (accessed 21 February 2023).

Noyes WA. Biographical Memoir of Julius Stieglitz 1867–1937. Natl. Acad. Sci. U. S. Am. Biogr. Mem., vol. XXI, Washington, D.C.: National Academy of Sciences, 1939, 273–314.

Nutrition and Alcoholism. J Am Med Assoc 1951; 146: 1618. https://doi.org/10.1001/jama.1951.03670170072026.

Nutton V, 2023: Galen. https://www.britannica.com/biography/Galen (accessed 21 February 2023).

O'Connell AN, Russo NF, editors. Anne Anastasi, 1908–. Models Achiev., Psychology Press, 1988.

OECD. OECD Glossary of Statistical Terms. OECD Publishing, 2008.

Office of the President at The University of Texas at Austin, 2023: Lorene Lane Rogers. https://president.utexas.edu/past-presidents/lorene-lane-rogers (accessed 21 February 2023).

Opie I, Opie P, editors. Jack Sprat. Oxf. Dict. Nurs. Rhymes, Oxford: Oxford University Press, 1977, 238.

Oregon State University Libraries, 2023: Special Collections & Archives Research Center. http://scarc.library.oregonstate.edu/ (accessed 25 July 2023).

Oregon State University Libraries Special Collections & Archives Research Center, 2014a: Roger J. Williams: Nutrition Scientist. https://paulingblog.wordpress.com/2014/06/26/roger-j-williams-nutrition-scientist/ (accessed 21 February 2023).

Oregon State University Libraries Special Collections & Archives Research Center, 2014b: Roger J. Williams and the Continuing Quest for Good Health.

https://paulingblog.wordpress.com/2014/07/02/roger-j-williams-and-the-continuing-quest-for-good-health/ (accessed 21 February 2023).

Orwell G. Animal Farm. 1st ed. New York: Harcourt, Brace and Company, 1946.

Paterson DG. A Public Relations Tool. Am Psychol 1954; 9: 642. https://doi.org/10.1037/h0063287.

Patterson AD, Turnbaugh PJ. Microbial Determinants of Biochemical Individuality and Their Impact on Toxicology and Pharmacology. Cell Metab 2014; 20: 761–768. https://doi.org/10.1016/j.cmet.2014.07.002.

Pauling L. Letter to Williams RJ, 7 January 1937. LP Correspondence 436.2 Williams, Roger J., 1937., Special Collections and Archives Research Center, Oregon State University Libraries and Press.

Pauling L. Letter to Williams RJ, 3 November 1957. LE Correspondence 436.10 Williams, Roger J., 1957, Special Collections and Archives Research Center, Oregon State University Libraries and Press.

Pennington D, Snell EE, Williams RJ. An Assay Method for Pantothenic Acid. J Biol Chem 1940; 135: 213–222.

Penrose LS. The Concept of Genetotrophic Disease. The Lancet 1950; 255: 464. https://doi.org/10.1016/S0140-6736(50)90386-2.

Perlman RL, Govindaraju DR. Archibald E. Garrod: The Father of Precision Medicine. Genet Med 2016; 18: 1088–1089. https://doi.org/10.1038/gim.2016.5.

Perreau K, 2023: BeReal. Your Friends for Real. https://bere.al/en (accessed 8 February 2023).

Pickren W, 2009: Anne Anastasi (1908-2001). https://www.apa.org/about/governance/president/anastasi (accessed 21 February 2023).

Pizzorno J. Roger J. Williams (1893-1988). Integr Med 2019; 18: 30–31.

Pollack MA, Taylor A, Taylor J, Williams RJ. B Vitamins in Cancerous Tissues I. Riboflavin. Cancer Res 1942a; 2: 739–743.

Pollack MA, Taylor A, Williams RJ. B Vitamins in Human, Rat and Mouse Neoplasms. Stud. Vitam. Content Tissues II, University of Texas Press, 1942b, 56–71, The University of Texas at Austin, Texas ScholarWorks.

Pollack MA, Taylor A, Woods A, Thompson RC, Williams RJ. B Vitamins in Cancerous Tissues III. Biotin. Cancer Res 1942c; 2: 748–751.

Pope A. Book Reviews. Psychosom Med 1958; 20: 171.

Powers RG. Not Without Honor: The History of American Anticommunism. New Haven: Yale University Press, 1998.

Procter BH, 2023: Sadler, McGruder Ellis. https://www.tshaonline.org/handbook/entries/sadler-mcgruder-ellis (accessed 21 February 2023).

Reader's Digest, 2023: About Reader's Digest. https://www.rd.com/about-readers-digest/ (accessed 21 February 2023).

Redfield R. The Human Frontier. Ecology 1947; 28: 212. https://doi.org/10.2307/1930959.

Reed J. A Study of Alcoholic Consumption and Amino Acid Excretion Patterns of Rats of Different Inbred Strains. Biochem. Inst. Stud. IV Individ. Metab. Patterns Hum. Dis. Explor. Study Util. Predominantly Pap. Chromatogr. Methods, Austin, Texas: The University of Texas, Austin, 1951a, 144–150.

Reed J. Individual Excretion Patterns in Laboratory Rats. Biochem. Inst. Stud. IV Individ. Metab. Patterns Hum. Dis. Explor. Study Util. Predominantly Pap. Chromatogr. Methods, Austin, Texas: The University of Texas, Austin, 1951b, 139–143.

Rees M, 1987: Warren Weaver 1894-1978 A Biographical Memoir. https://www.nasonline.org/wp-content/uploads/2024/06/weaver-warren.pdf. (accessed 22 January 2023).

Rheinberger H-J, Müller-Wille S. Vererbung: Geschichte und Kultur eines biologischen Konzepts. 1. Frankfurt am Main: FISCHER Taschenbuch, 2009.

Rhodes R. The Making of the Atomic Bomb. Trade Paperback Edition. London: Simon + Schuster UK, 2012.

Riddle O. Endocrines and Constitution in Doves and Pigeons. First Edition. Washington, D.C.: Carnegie Institute of Washington, 1947.

Rogers LL, Pelton RB, Williams RJ. Voluntary Alcohol Consumption by Rats Following Administration of Glutamine. J Biol Chem 1955; 214: 503–506.

Röhrig B, du Prel J-B, Wachtlin D, Kwiecien R, Blettner M. Sample size calculation in clinical trials: part 13 of a series on evaluation of scientific publications. Dtsch Arzteblatt Int 2010; 107: 552–556. https://doi.org/10.3238/arztebl.2010.0552.

Sackett DL, Rosenberg WM, Gray JA, Haynes RB, Richardson WS. Evidence Based Medicine: What It Is and What It Isn't. BMJ 1996; 312: 71–72. https://doi.org/10.1136/bmj.312.7023.71.

Salomon K, Jin A. Diathesis-Stress Model. In: Gellman MD, Turner JR, eds. Encycl. Behav. Med., New York, NY: Springer, 2013, 591–592. https://doi.org/10.1007/978-1-4419-1005-9_797.

Sanders R, 2003: Noted vitamin researcher Esmond Snell, former biochemistry chair at UC Berkeley, has died at 89. http://bioinst.cm.utexas.edu/Snell.htm (accessed 21 February 2023).

Schenarts PJ. Now Arriving: Surgical Trainees From Generation Z. J Surg Educ 2020; 77: 246–253. https://doi.org/10.1016/j.jsurg.2019.09.004.

Schloss JV. Nutritional deficiencies that may predispose to long COVID. Inflammopharmacology 2023; 31: 573–583. https://doi.org/10.1007/s10787-023-01183-3.

Schmidt-Erfurth U, Kohnen T, editors. Encyclopedia of Ophthalmology. 1st ed. Heidelberg, Germany: Springer-Verlag, 2018.

Shanks N, Greek R, Greek J. Are Animal Models Predictive for Humans? Philos Ethics Humanit Med PEHM 2009; 4: 2. https://doi.org/10.1186/1747-5341-4-2.

Shuman CR. Memoir of Michael G. Wohl 1889–1970. Trans Stud Coll Physicians Phila 1971; 38: 252–253.

Sigma Xi: The Scientific Research Honor Society, 2023: Robert R. Williams. https://www.sigmaxi.org/programs/prizes-awards/william-procter/award-winner/robert-r.-williams (accessed 21 February 2023).

Smith WD, 2022: Hippocrates. https://www.britannica.com/biography/Hippocrates (accessed 21 February 2023).

Smyrl VE, 2023: Moore, Harry Estill. https://www.tshaonline.org/handbook/entries/moore-harry-estill (accessed 21 February 2023).

Snell EE, Eakin RE, Williams RJ. A Quantitative Test for Biotin and Observations Regarding its Occurrence and Properties. J Am Chem Soc 1940a; 62: 175–178. https://doi.org/10.1021/ja01858a052.

Snell EE, Pennington D, Williams RJ. The Effect of Diet on the Pantothenic Acid Content of Chick Tissues. J Biol Chem 1940b; 133: 559–565.

Snell EE, Williams RJ. Biotin as a Growth Factor for the Butyl Alcohol Producing Anaerobes. J Am Chem Soc 1939; 61: 3594. https://doi.org/10.1021/ja01267a515.

Spies TD, Stanbery SR, Williams RJ, Jukes TH, Babcock SH. Pantothenic Acid in Human Nutrition. J Am Med Assoc 1940; 115: 523–524. https://doi.org/10.1001/jama.1940.72810330004010b.

Srb AM. G. W. Beadle. Annu Rev Genet 1990; 24: 1–6. https://doi.org/10.1146/annurev.ge.24.120190.000245.

SRI International, 2014: HumanCyc: Homo Sapiens Myo-inositol Biosynthesis. https://biocyc.org/HUMAN/NEW-IMAGE?type=PATHWAY&object=PWY-2301 (accessed 21 February 2023).

SRI International, 2023: HumanCyc: Cellular Overview. https://biocyc.org/overviewsWeb/celOv.shtml?orgid=HUMAN&pnids=PWY-2301 (accessed 21 February 2023).

Steg A, Oczkowicz M, Smołucha G. Omics as a Tool to Help Determine the Effectiveness of Supplements. Nutrients 2022; 14: 5305. https://doi.org/10.3390/nu14245305.

Stevenson A, editor. Negro. Oxf. Dict. Engl. 3. Edition, Oxford: Oxford University Press, 2010.

Stueck W. The Korean War: An International History. Princeton, N.J: Princeton University Press, 1995.

Tawa EA, Hall SD, Lohoff FW. Overview of the Genetics of Alcohol Use Disorder. Alcohol Alcohol 2016; 51: 507–514. https://doi.org/10.1093/alcalc/agw046.

Taylor A, Carmichael N, Young Jr. MK. Cancer studies II. University of Texas, 1953.

Taylor A, Pollack MA, Hofer MJ, Williams RJ. B Vitamins in Cancerous Tissues II. Nicotinic Acid. Cancer Res 1942a; 2: 744–747.

Taylor A, Pollack MA, Hofer MJ, Williams RJ. B Vitamins in Cancerous Tissues IV. Pantothenic Acid. Cancer Res 1942b; 2: 752–754.

Taylor A, Pollack MA, Williams RJ. Uniformities in the Content of B Vitamins in Malignant Neoplasms. Science 1942c; 96: 322–323. https://doi.org/10.1126/science.96.2492.322-a.

Taylor A, Williams RJ. Diet and Spontaneous Lung Tumors in Strain A Mice. Cancer Stud., Austin, Texas: The University of Texas, Austin, 1945, 119–122.

Templin MC. John E. Anderson: 1893-1966. Child Dev 1968; 39: 657–670.

The Eastern Pennsylvania Branch of the American Society for Microbiology, 1987: L. Joe Berry, PhD. Obituary. https://epaasm.org/obituaries/l-joe-berry-phd-obituary-february-26-1987/ (accessed 21 February 2023).

The Editors. Brief Comment on Recent Books. Am J Psychol 1949; 62: 148–158.

The Editors. Alcoholism. Eng Sci 1952; XV: 9–12, Caltech Library.

The Editors of Encyclopaedia Britannica, 2022: Robert Redfield. https://www.britannica.com/biography/Robert-Redfield (accessed 21 February 2023).

The Human Frontier: A New Pathway for Science Toward a Better Understanding of Ourselves. J Am Med Assoc 1947; 134: 741. https://doi.org/10.1001/jama.1947.02880250089030.

The New York Times, 1940: Vital Acid of Life Now Synthesized. https://www.nytimes.com/1940/03/08/archives/vital-acid-of-life-now-synthesized-production-of-vitamin-needed-by.html (accessed 4 April 2021).

The New York Times, 1942: Morality is Linked to Vitamins in Diet. https://www.nytimes.com/1942/02/27/archives/morality-is-linked-to-vitamins-in-diet-dr-rj-williams-says-tests.html (accessed 4 April 2021).

The New York Times, 1952: Scientist's Wife Killed. https://www.nytimes.com/1952/02/17/archives/scientists-wife-killed-mrs-roger-j-williams-steps-into-trains-path.html (accessed 4 April 2021).

The New York Times, 1975: Dr. Edward W. Dempsey Dies; Anatomy Professor at Columbia (12.01.1975). https://www.nytimes.com/1975/01/12/archives/dr-edward-w-dempsey-dies-anatomy-professor-at-columbia.html (accessed 22 October 2020).

The Philosophical Society of Texas. Proceedings of The Annual Meeting of the Philosophical Society of Texas. Austin, Texas: The Philosophical Society of Texas, 1949.

Thompson RC, Kirby HM. Biochemical Individuality II: Variation in the Urinary Excretion of Lysine, Threonine, Leucine, and Arginine. Arch Biochem 1949; 21: 210–216.

Tinsley JA, 2023: Clayton, William Lockhart. https://www.tshaonline.org/handbook/entries/clayton-william-lockhart (accessed 21 February 2023).

UConn Foundation, 2023: Randolph T. Major Lecture Series. http://na.eventscloud.com/ehome/406223/858155/ (accessed 21 February 2023).

Uetrecht J. Idiosyncratic Drug Reactions: Past, Present, and Future. Chem Res Toxicol 2008; 21: 84–92. https://doi.org/10.1021/tx700186p.

Uetrecht J, Naisbitt DJ. Idiosyncratic Adverse Drug Reactions: Current Concepts. Pharmacol Rev 2013; 65: 779–808. https://doi.org/10.1124/pr.113.007450.

Universitäts- und Landesbibliothek Bonn, 2023: BONNUS. https://bonnus.ulb.uni-bonn.de/ (accessed 25 July 2023).

US Census Bureau, 1953: Statistical Abstract of the United States: 1953. https://www.census.gov/library/publications/1953/compendia/statab/74ed.html (accessed 27 February 2023).

U.S. Department of Labor, 2023: Individualized Learning Plan. http://www.dol.gov/agencies/odep/program-areas/individuals/youth/individualized-learning-plan (accessed 3 March 2023).

Visvikis-Siest S, Theodoridou D, Kontoe M-S, Kumar S, Marschler M. Milestones in Personalized Medicine: From the Ancient Time to Nowadays – The Provocation of COVID-19. Front Genet 2020; 11: 1–12.

Vogel F. II. Moderne Probleme der Humangenetik. Ergeb. Inn. Med. Kinderheilkd., Heidelberg, Germany: Springer, 1959, 52–125.

Walter J. Murphy. Chem Eng News Arch 1959; 37: 7. https://doi.org/10.1021/cen-v037n049.p007.

Warren E, Supreme Court of The United States, 1954: U.S. Reports: Brown v. Board of Education, 347 U.S. 483 (1954). https://www.loc.gov/item/usrep347483/ (accessed 21 February 2023).

Watson DJ. The Double Helix. London: W&N, 2010.

Watson HE. A Century of Book Publishing, 1848–1948. Nature 1949; 163: 659–660. https://doi.org/10.1038/163659a0.

Watson TJ. Introducing "THINK". Think 1935; 1: 3.

Westad OA. The Cold War: A World History. New York: Basic Books, 2017.

What to Do About Vitamins. J Am Med Assoc 1945; 129: 241. https://doi.org/10.1001/jama.1945.02860370063036.

Wieman HN, Peden C. The Life and Thought of Henry Nelson Wieman (1884–1975): An American Philosopher. Lewiston, NY: Edwin Mellen Press, 2010.

Wildiers E. Nouvelle substance indispensable au développement de la lévure. La Cellule 1901; 18: 313–332.

Wiley, 2023: Corporate Information. https://www.wiley.com/WileyCDA/Section/id-301454.html (accessed 2 March 2023).

Williams RJ. The Vitamine Requirement of Yeast: A Simple Biological Test for Vitamine. J Biol Chem 1919; 38: 465–486, Journal of Biological Chemistry Archive.

Williams RJ. A Quantitative Method for Determination of Vitamine. J Biol Chem 1920; 42: 259–265.

Williams RJ. "Taste Deficiency" for Creatine. Science 1931; 74: 597–598. https://doi.org/10.1126/science.74.1928.597.

Williams RJ. An Introduction to Organic Chemistry. 3rd ed. New York: D. Van Nostrand Company, Inc., 1935a.

Williams RJ. Amino Acids, Proteins and Related Substances. Introd. Org. Chem. 3rd ed., New York: D. Van Nostrand Company, Inc., 1935b, 284–315.

Williams RJ. Organic Substances of Special Biochemical Interest. Introd. Org. Chem. 3rd ed., New York: D. Van Nostrand Company, Inc., 1935c, 559–566.

Williams RJ. Letter to Pauling L, 27 December 1936. LP Correspondence 436.1 Williams, Roger J., 1936., Special Collections and Archives Research Center, Oregon State University Libraries and Press.

Williams RJ. Letter to Pauling L, 13 December 1937. LP Correspondence 436.2 Williams, Roger J., 1937., Special Collections and Archives Research Center, Oregon State University Libraries and Press.

Williams RJ. M(anille) Ide, the Discoverer of "Bios". Science 1938; 88: 475. https://doi.org/10.1126/science.88.2290.475-b.

Williams RJ. Pantothenic Acid – A Vitamin. Science 1939a; 89: 486. https://doi.org/10.1126/science.89.2317.486.

Williams RJ. Letter to Pauling L, 30 January 1939b. LP Correspondence, 436.4 Williams, Roger J., 1939, Special Collections and Archives Research Center, Oregon State University Libraries and Press.

Williams RJ. Vitamin Study at the University of Texas. Science 1940; 92: 579. https://doi.org/10.1126/science.92.2399.579-a.

Williams RJ. The Importance of Microorganisms in Vitamin Research. Science 1941a; 93: 412–414. https://doi.org/10.1126/science.93.2418.412.

Williams RJ. Growth-Promoting Nutrilites for Yeasts. Biol Rev 1941b; 16: 49–80. https://doi.org/10.1111/j.1469-185X.1941.tb01095.x.

Williams RJ. Introduction. Stud. Vitam. Content Tissues I, University of Texas Press, 1941c, The University of Texas at Austin, Texas ScholarWorks.

Williams RJ. Vitamins in the Future. Science 1942a; 95: 340–344. https://doi.org/10.1126/science.95.2466.340.

Williams RJ. The Approximate Vitamin Requirements of Human Beings. J Am Med Assoc 1942b; 119: 1–6.

Williams RJ. The Significance of the Vitamin Content of Tissues. In: Harris RS, Thimann KV, McCollum EV, eds. Vitam. Horm., vol. 1, Academic Press, 1943a, 229–247. https://doi.org/10.1016/S0083-6729(08)60256-3.

Williams RJ. The Chemistry and Biochemistry of Pantothenic Acid. In: Nord FF, Werkman CH, eds. Adv. Enzymol. Relat. Areas Mol. Biol., New York: John Wiley & Sons, 1943b, 253–287. https://doi.org/10.1002/9780470122488.ch8.

Williams RJ. Pantothenic Acid and the Microbiological Approach to the Study of Vitamins. Biol Action Vitam Symp 1944: 120–135.

Williams RJ. What To Do About Vitamins. Norman, Oklahoma: The University of Oklahoma Press, 1945a.

Williams RJ. "Starring" in American Men of Science in Relation to the Status of Chemistry. Science 1945b; 101: 222–223. https://doi.org/10.1126/science.101.2618.222.

Williams RJ. The Clinical Possibilities of Pantothenic Acid. Dietotherapy Clin Appl Mod Nutr 1945d: 263–267, 288–289.

Williams RJ. Achieving Full Employment After the War. Science 1945e; 101: 537. https://doi.org/10.1126/science.101.2630.537.

Williams RJ. The Human Frontier: A New Pathway for Science Toward a Better Understanding of Ourselves. New York: Harcourt, Brace and Company, 1946a.

Williams RJ. Production of Pantothenic Acid and Other Related Growth Promoting Substances. US2414682A, 1947a.

Williams RJ. Humanics: A Crucial Need. Sci Mon 1947b; 64: 174–180.

Williams RJ. Will Science Meet a New Challenge? Am Sci 1947d; 35: 282–286.

Williams RJ. The Etiology of Alcoholism: A Working Hypothesis Involving the Interplay of Hereditary and Environmental Factors. Q J Stud Alcohol 1947e; 7: 567–587.

Williams RJ. Biochemical Individuality and Its Implications. Chem Eng News 1947j; 25: 1112.

Williams RJ. We're All Peculiar. Read Dig Mult Lang 1948a: 17–19.

Williams RJ. Why People Are Different. Think 1948b; 14: 10, 24.

Williams RJ. Alcoholics and Metabolism. Sci Am 1948c; 179: 50–53.

Williams RJ. Biochemical Approach to Individuality. Science 1948d; 107: 459. https://doi.org/10.1126/science.107.2784.457.

Williams RJ. Shall WE Pioneer Too? The Annual Meeting of the Philosophical Society of Texas: Austin, Texas, 1949b.

Williams RJ. Some Implications of Physiological Individuality. Feel Emot 1950b: 268–273.

Williams RJ. Concept of Genetotrophic Disease. Nutr Rev 1950c; 8: 257–260. https://doi.org/10.1111/j.1753-4887.1950.tb02469.x.

Williams RJ. Nutrition and Alcoholism. Norman, Oklahoma: University of Oklahoma Press, 1951a.

Williams RJ. Introduction, General Discussion and Tentative Conclusions. Biochem. Inst. Stud. IV Individ. Metab. Patterns Hum. Dis. Explor. Study Util. Predominantly Pap. Chromatogr. Methods, Austin, Texas: The University of Texas, Austin, 1951b, 7–21.

Williams RJ. The Unexplored Field of Genetotrophic Disease. M D 1951c; 6: 123, 124, 136.

Williams RJ. Men and Marbles. San Antonio: Sixth Southwestern Regional Meeting of the American Chemical Society, 1951d.

Williams RJ. Letters. Sci Am 1951e; 185: 4.

Williams RJ. Nutritional Vulnerability and Alcoholism. Res Rev Publ Off Nav Res 1952a: 4–8.

Williams RJ. Alcoholism as a Nutritional Problem. Am J Clin Nutr 1952b; 1: 32–36. https://doi.org/10.1093/ajcn/1.1.32.

Williams RJ. Free and Unequal: The Biological Basis of Individual Liberty. Austin, Texas: University of Texas Press, 1953a.

Williams RJ. Muscular Dystrophy and Individual Metabolic Patterns: The Possibilities of a Nutritional Therapeutic Approach. Proc. First Second Med. Conf. 1951–1952 Muscular Dystrophy Assoc. Am. Inc, 1953c, 118–122.

Williams RJ. Book Review: Advances in Cancer Research, Volume I: Edited by Jesse P. Greenstein, National Cancer Institute, Bethesda, Maryland. Academic Press Inc., New York, N. Y., 1953. xii + 590 pp. Price $12.00. Arch Biochem Biophys 1953d; 47: 233–234. https://doi.org/10.1016/0003-9861(53)90460-3.

Williams RJ, 1954a: Autobiography. http://bioinst.cm.utexas.edu/williams/Autobiogr.htm (accessed 21 February 2023).

Williams RJ. Early Experiences with Pantothenic Acid – A Retrospect. Nutr Rev 1954b; 12: 65–68. https://doi.org/10.1111/j.1753-4887.1954.tb03197.x.

Williams RJ. The Genetotrophic Concept – Nutritional Deficiencies and Alcoholism. Ann N Y Acad Sci 1954d; 57: 794–811. https://doi.org/10.1111/j.1749-6632.1954.tb36457.x.

Williams RJ. Implications of Humanics for Law and Science. J Public Law 1954e; 3: 328–344.

Williams RJ. The Genetotrophic Approach to Alcoholism. Orig. Resist. Toxic Agents, Washington, D.C.: Academic Press, N.Y., 1954f, 194–208.

Williams RJ. Biochemical Approach to the Study of Personality. Psychiatr. Res. Rep. Am. Psychiatr. Assoc., American Psychiatric Association, 1954g, 31–33.

Williams RJ. Chemical Anthropology: Alcoholism (presented to Am. Chem. Soc., Dec. 13, 1954). The Vortex 1955a; 16: 68–80.

Williams RJ. Biochemical Individuality: The Basis for the Genetotrophic Concept. 1st ed. New York: John Wiley and Sons, 1956a.

Williams RJ. The Genetotrophic Approach. Biochem. Individ. Basis Genet. Concept. 1st ed., New York: John Wiley and Sons, 1956b, 166–176.

Williams RJ. Normal Young Men. Perspect Biol Med 1957a; 1: 97–104. https://doi.org/10.1353/pbm.1957.0001.

Williams RJ. Forty Ways To Be Dumb. J Chem Educ 1957b; 34: 261. https://doi.org/10.1021/ed034p261.

Williams RJ. Individuality and Education. Educ Leadersh J Assoc Superv Curric 1957c; 14: 144–148.

Williams RJ. Nutrition in a Nutshell. First. New York: Dolphin Books, 1962.

Williams RJ. Free and Unequal: The Biological Basis of Individual Liberty. New York: John Wiley & Sons, 1964.

Williams RJ. How Can the Climate in Medical Education Be Changed? Perspect Biol Med 1971; 14: 608–614. https://doi.org/10.1353/pbm.1971.0039.

Williams RJ. A Flaw in Medical Education. Nutr Today 1972; May/June: 30–33.

Williams RJ. The Wonderful World Within You: Your Inner Nutritional Environment. New York City: Bantam Books, 1977.

Williams RJ. Free and Unequal: The Biological Basis of Individual Liberty. Indianapolis: Liberty Fund, 1979.

Williams RJ. Rethinking Education: The Coming Age of Enlightenment. Philosophical Library, 1986.

Williams RJ. Biochemical Individuality: The Basis for the Genetotrophic Concept. New Canaan, Conneticut: Keats Publishing, Inc., 1998.

Williams RJ. The Human Frontier. Franklin Classics, 2018.

Williams RJ, Beerstecher E Jr, Eldon Sutton H, Kirby HK, Brown WD, Reed J, Rich GB, Berry LJ. Biochemical Individuality. V: Explorations with Respect to the Metabolic Patterns of Compulsive Drinkers. Arch Biochem 1950a; 29: 27–40.

Williams RJ, Beerstecher Jr. E. An Introduction to Biochemistry. 1st ed. New York, Toronto, London: D. Van Nostrand Company, Inc., 1931.

Williams RJ, Beerstecher Jr. E. An Introduction to Biochemistry. 2nd ed. New York, Toronto, London: D. Van Nostrand Company, Inc., 1948a.

Williams RJ, Beerstecher Jr. E. The Nutritional Requirements of Mammals. Introd. Biochem. 2nd ed., New York, Toronto, London: D. Van Nostrand Company, Inc., 1948b, 229–290.

Williams RJ, Beerstecher Jr. E, Joe Berry L. The Concept of Genetotrophic Disease. The Lancet 1950c; 255: 287–289. https://doi.org/10.1016/S0140-6736(50)91997-0.

Williams RJ, Berry LJ, Beerstecher Jr. E. Biochemical Individuality III. Genetotrophic Factors in the Etiology of Alcoholism. Arch Biochem 1949a; 23: 275–290.

Williams RJ, Berry LJ, Beerstecher Jr. E. Individual Metabolic Patterns, Alcoholism, Genetotrophic Diseases. Proc Natl Acad Sci U S A 1949b; 35: 265–271.

Williams RJ, Berry LJ, Beerstecher Jr. E. Genetotrophic Diseases; Alcoholism. Tex Rep Biol Med 1950d; 8: 238–256.

Williams RJ, Bradway EM. The Further Fractionation of Yeast Nutrilites and Their Relationship to Vitamin B and Wildiers' "Bios". J Am Chem Soc 1931; 53: 783–789. https://doi.org/10.1021/ja01353a051.

Williams RJ, Brown WD, Shideler RW. Metabolic Peculiarities in Normal Young Men as Revealed by Repeated Blood Analyses. Proc Natl Acad Sci U S A 1955a; 41: 615–620.

Williams RJ, Cheldelin VH, Eppright MA, Snell EE, Guirard BM, Mitchell HK, Isbell ER, Taylor A, Pollack MA, Sortomme CL, Woods AM, Taylor J, Hofer MJ, Johnson GA, Lane RL, McMahan JR, Thompson RC, Pennington D, Thacker J. Studies on the Vitamin Content of Tissues II. Austin, Texas: University of Texas Press, 1942, The University of Texas at Austin, Texas ScholarWorks.

Williams RJ, Davis DR, Hackert ML, 1966: The Clayton Foundation Biochemical Institute: A Short History. https://bioinst.cm.utexas.edu/History_RJW.pdf (accessed 23 February 2023).

Williams RJ, Eakin RE, Beerstecher Jr. E, Shive W. The Biochemistry of B Vitamins. New York: Reinhold Publishing Corporation, 1950e.

Williams RJ, Eakin RE, Snell EE. The Relationship of Inositol, Thiamin, Biotin, Pantothenic Acid and Vitamin B6 to the Growth of Yeasts. J Am Chem Soc 1940a; 62: 1204–1207. https://doi.org/10.1021/ja01862a062.

Williams RJ, Honn JM. Role of "Nutrilites" in the Nutrition of Molds and Other Fungi. Plant Physiol 1932; 7: 629–641. https://doi.org/10.1104/pp.7.4.629.

Williams RJ, Kirby Berry H, Sutton HE, Cain L, Berry JS, Jirgendsons B, Bloch E, Beerstecher Jr. E, Thompson RC, Reed J, Brown WD, Berry LJ, Rogers LL, Brown JD, Young Jr. JD. Biochemical Institute Studies IV: Individual Metabolic Patterns and Human Disease: An Exploratory Study Utilizing Predominantly Paper Chromatographic Methods. Austin, Texas: The University of Texas, Austin, 1951.

Williams RJ, Lasselle PA. The Identification of Creatine. J Am Chem Soc 1926; 48: 536–537. https://doi.org/10.1021/ja01413a036.

Williams RJ, Lyman CM, Goodyear GH, Truesdail JH, Holaday D. "Pantothenic Acid," A Growth Determinant of Universal Biological Occurrence. J Am Chem Soc 1933; 55: 2912–2927. https://doi.org/10.1021/ja01334a049.

Williams RJ, Major RT. The Structure of Pantothenic Acid. Science 1940; 91: 246. https://doi.org/10.1126/science.91.2358.246.

Williams RJ, Mitchell HK, Weinstock HH, Snell EE. Pantothenic Acid. VII. Partial and Total Synthesis Studies. J Am Chem Soc 1940b; 62: 1784–1785. https://doi.org/10.1021/ja01864a038.

Williams RJ, Pelton RB, Rogers LL. Dietary Deficiencies in Animals in Relation to Voluntary Alcohol and Sugar Consumption. Q J Stud Alcohol 1955b; 16: 234–244.

Williams RJ, Rogers LL. The Formulation of a Genetotrophic Supplement for the Experimental Treatment of Diseases of Obscure Etiology. Tex Rep Biol Med 1953; 11: 573–578.

Williams RJ, Saunders DH. The Effects of Inositol, Crystalline Vitamin B1 and "Pantothenic Acid" on the Growth of Different Strains of Yeast. Biochem J 1934; 28: 1887–1893. https://doi.org/10.1042/bj0281887.

Williams RJ, Schlenk F, Eppright MA. The Assay of Purified Proteins, Enzymes, Etc., for "B Vitamins". J Am Chem Soc 1944; 66: 896–898. https://doi.org/10.1021/ja01234a016.

Williams RJ, Strong FM, Mitchell HK, McMahan JR, Eakin RE, Wright LD, Snell EE, Stout AK, Cheldelin VH, Taylor A, Pollack MA. Studies on the Vitamin Content of Tissues I. Austin, Texas: University of Texas Press, 1941, The University of Texas at Austin, Texas ScholarWorks.

Williams RJ, Taylor A, Kynette A, Hungate RE, Thompson RC, Morgan BB, Snider H, Loo YH. Cancer Studies. Austin, Texas: The University of Texas, Austin, 1945.

Williams RJ, Truesdail JH. The Use of Fractional Electrolysis in the Fractionation of the "Bios" of Wildiers. J Am Chem Soc 1931; 53: 4171–4181. https://doi.org/10.1021/ja01362a036.

Williams RJ, Warner ME, Roehm RR. The Effect of Various Preparations on the Growth of Bakers' and Brewers' Yeasts. J Am Chem Soc 1929; 51: 2764–2774. https://doi.org/10.1021/ja01384a024.

Williams RJ, Wilson JL, Von der Ahe FH. The Control of "Bios" Testing and the Concentration of a "Bios". J Am Chem Soc 1927; 49: 227–235. https://doi.org/10.1021/ja01400a026.

Williams RR. Natural Science and Social Problems. Am Sci 1948e; 36: 116–126.

Williams RR. Toward the Conquest of Beriberi. Cambridge, Massachusetts: Harvard University Press, 2013.

Woodworth RS. Review of Free and unequal: The biological basis of individual liberty. J Appl Psychol 1955; 39: 67. https://doi.org/10.1037/h0038825.

World Health Organization, 2018: Global Status Report on Alcohol and Health 2018. https://www.who.int/publications-detail-redirect/9789241565639 (accessed 5 March 2023).

WorldCat, 2023: Garside, E. B. (Edward Ballard) 1907–1999. http://worldcat.org/identities/lccn-no96046272/ (accessed 21 February 2023).

Wright LD, McMahan JR, Cheldelin VH, Taylor A, Snell EE, Williams RJ, Snell EE, Stout AK, Cheldelin VH, Taylor A, Pollack MA. The "B Vitamins" in Normal Tissues (Autolysates). Stud. Vitam. Content Tissues I, Austin, Texas: University of Texas Press, 1941, 38–60, The University of Texas at Austin, Texas ScholarWorks.

Yee K, 2020: L. Joe Berry Papers. https://lib.guides.umbc.edu/c.php?g=836720&p=6572811 (accessed 8 March 2021).

Zimmerman HJ. Various forms of chemically induced liver injury and their detection by diagnostic procedures. Environ Health Perspect 1976; 15: 3–12. https://doi.org/10.1289/ehp.76153.

20 Index

Archibald Garrod, 11, 25, 46, 93, 94, 97, 128, 156, 226
Arno Motulsky, 13

Barry J. Anson, 122, 195, 204, 271, 286
Beadle and Tatum, 25, 92, 226
Benjamin Clayton, 32, 63, 143, 238

Clark W. Heath, 102, 116

Earnest A. Hooton, 104
Ernest Beerstecher Jr., 154

Fred Conrad Koch, 31, 281

George Draper, 95, 118

Julius Stieglitz, 31, 288

Karl Lashley, 196, 236

L. Joe Berry, 124, 138, 154, 213, 214, 294, 306
Leo Loeb, 94, 116, 280
Linus Pauling, 32, 52, 238, 249
Lionel S. Penrose, 156
Lorene L. Rogers, 193, 213, 214

McGruder Ellis Sadler, 142
Mitchell and Houlahan, 150

Robert Fleming, 129
Robert R. Williams, 31, 57, 105, 238, 292

Theodor Brugsch, 15

Walter J. Murphy, 163, 281, 296
Warren Weaver, 28, 219, 238, 254, 265, 266, 268, 269, 291
William Lockhart Clayton, 143

21 Acknowledgments

This dissertation is the culmination of five years of research and has been significantly influenced by numerous individuals, all of whom I am greatly indebted to. I would like to thank Prof. Dr. phil. Dr. rer. med. habil. Mariacarla Gadebusch Bondio for not only inspiring the research topic this work covers, but her continued support through the entire process from its inception to final completion. Her insightful critiques and comments have been invaluable to the completion of this work. I am additionally grateful to Julia Letow, Dr. Felix Sommer, Dr. Christian Kaiser, Dr. Sarah Diner, and the entire Institute for Medical Humanities of the University of Bonn; their efforts in educating us medical students in the intricacies involved in historical academic work, their advice, and their insight into the research process was crucial from the outset of this undertaking.

Donald R. Davis has been an instrumental source of reading material, knowledge, and comprehension, accompanying my research process for many years and providing countless original sources which would otherwise have been impossible to obtain. His readiness for spirited discourse, as well as his availability for letters of recommendation and an interview, have been remarkable. I would additionally like to thank Margaret Schlankey and the entire team of the Dolph Briscoe Center for American History at The University of Texas at Austin for their help in the organisation and execution of a one-month research trip to Austin, as well as the financial support granted by The William and Madeline Welder Smith Travel Award.

The outstanding influence of Leonie Brand's linguistic prowess, coupled with her extensive caseological insights and untiring efforts in finding even the smallest typographical errors, cannot be understated. I thank her and Katharina Heller for their patience, understanding, and unwavering support.

Medizingeschichte im Kontext

Herausgegeben von Karl-Heinz Leven, Mariacarla Gadebusch Bondio,
Hans-Georg Hofer und Livia Prüll

Die Reihe *Medizingeschichte im Kontext* veröffentlicht Studien, die Themen aus der Geschichte der Medizin und des Gesundheitswesens in wissenschafts- und kulturhistorischer Perspektive betrachten. Die Reihe versteht sich zugleich als Fortsetzung der von Ludwig Aschoff 1938/39 mit zwei Heften begründeten, von Eduard Seidler 1971-1994 mit 17 Bänden weitergeführten *Freiburger Forschungen zur Medizingeschichte*. Die Bände 1 bis 11 (1999 bis 2004) wurden von Karl-Heinz Leven und Ulrich Tröhler herausgegeben.

Band 1 Christine Hummel: Das Kind und seine Krankheiten in der griechischen Medizin. Von Aretaios bis Johannes Aktuarios (1. bis 14. Jahrhundert). 1999.

Band 2 Cécile Mack: Henriette Hirschfeld-Tiburtius (1834-1911). Das Leben der ersten selbständigen Zahnärztin Deutschlands. 1999.

Band 3 Susanne Mende: Die Wiener Heil- und Pflegeanstalt *Am Steinhof* im Nationalsozialismus. 2000.

Band 4 Bernhard Gessler: Eugen Fischer (1874-1967). Leben und Werk des Freiburger Anatomen, Anthropologen und Rassenhygienikers bis 1927. 2000.

Band 5 Jochen Binder: Zwischen Standesrecht und Marktwirtschaft. Ärztliche Werbung zu Beginn des 20. Jahrhunderts im deutsch-englischen Vergleich. 2000.

Band 6 Cécile Mack: Die badische Ärzteschaft im Nationalsozialismus. 2001.

Band 7 Beate Waigand: Antisemitismus auf Abruf. Das Deutsche Ärzteblatt und die jüdischen Mediziner 1918-1933. 2001.

Band 8 Georg Schomerus: Ein Ideal und sein Nutzen. Ärztliche Ethik in England und Deutschland 1902-1933. 2001.

Band 9 Barbara Rabi: Ärztliche Ethik – Eine Frage der Ehre? Die Prozesse und Urteile der ärztlichen Ehrengerichtshöfe in Preußen und Sachsen 1918-1933. 2002.

Band 10 Bernd Grün / Hans-Georg Hofer / Karl-Heinz Leven (Hrsg.): Medizin und Nationalsozialismus. Die Freiburger Medizinische Fakultät und das Klinikum in der Weimarer Republik und im „Dritten Reich". 2002.

Band 11 E. Caroline Jagella: Ignaz Schwörer (1800–1860). Freiburger Geburtshelfer zwischen Romantik und Positivismus. Ein Beitrag zur Geschichte der medizinischen Ethik im 19. Jahrhundert. 2004.

Band 12 Stephan Anis Towfigh: Das Bahá'ítum und die Medizin. Ein medizinhistorischer Beitrag zum Verhältnis von Religion und Medizin. 2006.

Band 13 Nils Kessel: Geschichte des Rettungsdienstes 1945–1990. Vom „Volk von Lebensrettern" zum Berufsbild „Rettungsassistent/in". 2008.

Band 14 Jette Sophia Jung: Erfolg und Scheitern der Hegar-Operation. Eine wissenschaftsgeschichtliche Untersuchung über die Kastration der Frau im 19. Jahrhundert. 2007.

Band 15 Jasmin Beatrix Mattes: Die Stationsbenennungen des Klinikums der Albert-Ludwigs-Universität Freiburg im Breisgau. Erinnerungskultur, kollektives Gedächtnis und Umgang mit nationalsozialistischer Vergangenheit. 2008.

Band 16 Simon Reuter: Im Schatten von Tet. Die Vietnam-Mission der Medizinischen Fakultät Freiburg (1961–1968). 2011.

Band 17 Ute Caumanns / Fritz Dross / Anita Magowska (Hrsg. / red.): Medizin und Krieg in historischer Perspektive. Beiträge der XII. Tagung der Deutsch-Polnischen Gesellschaft für Ge- schichte der Medizin, Düsseldorf 18.-20. September 2009. Medycyna i wojna w perspek- tywie historycznej. Prace XII. konferencji Polsko-Niemieckiego Towarzystwa Historii Medycyny, Düsseldorf 18 do 20 września 2009 r.. 2012.

Band 18 Philipp Rauh / Karl-Heinz Leven: Ernst Wilhelm Baader (1892-1962) und die Arbeitsmedizin im Nationalsozialismus. 2013.

Band 19 Eva Brinkschulte / Mariacarla Gadebusch Bondio (Hrsg.): Norm als Zwang, Pflicht und Traum. Normierende versus individualisierende Bestrebungen in der Medizin. Festschrift zum 60. Geburtstag von Heinz-Peter Schmiedebach. 2015.

Band 20 Eva Brinkschulte / Fritz Dross / Anita Magowska / Marcin Moskalewicz / Philipp Teichfi- scher (Hrsg./red.): Medizin und Sprache – Die Sprache der Medizin. Medycyna i język – język medycyny. 2016.

Band 21 Jessica Tannenbaum: Medizin im Konzentrationslager Flossenbürg 1938 bis 1945. Biografische Annäherungen an Täter, Opfer und Tatbestände. 2017.

Band 22 Simone Kahlow: Archäologie des Hospitals. *Pauperes et infirmi* in Fürsorgeinstitutionen nördlich der Alpen vom 12. bis zum 19. Jahrhundert. 2020.

Band 23 Dana Derichs: Die Medizinstudentinnen der Universität Erlangen in der Weimarer Republik und im Nationalsozialismus. 2022.

Band 24 Robert Davidson: Gustav Kolb und die Reformpsychiatrie in Erlangen 1911–1934. 2022.

Band 25 Jan Esse: Malaria in Südwest-Afrika Deutsche Kolonialmedizin 1884–1915. 2022.

Band 26 Georg-Benedict Brand: The Origins and Development of Roger J. Williams' Concept of Biochemical Individuality. 2024.

www.peterlang.com

www.ingramcontent.com/pod-product-compliance
Ingram Content Group UK Ltd.
Pitfield, Milton Keynes, MK11 3LW, UK
UKHW021822140426
5217IPUK00004B/41